Surround Sound

Surround Sound:
Up and Running

Second Edition

Tomlinson Holman

AMSTERDAM • BOSTON • HEIDLEBERG • LONDON
NEWYORK • OXFORD • PARIS • SAN DIEGO
SAN FRANCISCO • SINGAPORE • SYDNEY • TOKYO
Focal Press is an imprint of Elsevier

Acquisitions Editor: Catharine Steers
Publishing Services Manager: George Morrison
Project Manager: Mónica González de Mendoza
Assistant Editor: David Bowers
Marketing Manager: Christine Degon Veroulis
Cover Design: Joanne Blank

Focal Press is an imprint of Elsevier
30 Corporate Drive, Suite 400, Burlington, MA 01803, USA
Linacre House, Jordan Hill, Oxford OX2 8DP, UK

Library of Congress Cataloging-in-Publication Data
Holman, Tomlinson.
Surround sound up and running / Tomlinson Holman.—2nd ed.
p. cm.
Includes index.
ISBN 978-0-240-80829-1 (pbk. : alk. paper) 1. Surround-sound systems. I. Title.
TK7881.83.H65 2007
621.389'3--dc22

2007031553

British Library Cataloguing-in-Publication Data
A catalogue record for this book is available from the British Library.

ISBN: 978-0-240-80829-1

For information on all Focal Press publications
visit our website at www.books.elsevier.com

07 08 09 10 11 12 13 10 9 8 7 6 5 4 3 2 1

Typeset by Charon Tec Ltd (A Macmillan Company), Chennai, India
www.charontec.com

Printed in the United States of America

Frontispiece

This work is dedicated to the memory of John M. Eargle, mentor, colleague, and friend.

Table of Contents

Preface to the Second Edition xi

1 Introduction 1
A Brief History 2

2 Monitoring 23
Introduction 24
How Monitoring Affects the Mix 25
Full-Range Monitoring 25
Spatial Balance 26
Room Acoustics for Multichannel Sound 28
Choice of Monitor Loudspeakers 33
One Standardized Setup 36
Left and Right 36
Surround 37
Subwoofer 38
Setting Up the Loudspeaker Locations with Two Pieces of String 38
Setup Compromises 40
Center 40
Left and Right 42
Surround 44
Subwoofer 44
Setup Variations 46
Use of Surround Arrays 46
Surround Loudspeaker Directivity 48
Square Array 49
Close-Field Monitoring 50

Time Adjustment of the Loudspeaker Feeds 52
Low-Frequency Enhancement—The 0.1 Channel 53
Film Roots 53
Headroom on the Medium 54
Digital Film Sound Enters the Picture 56
Bass Management or Redirection 56
Digital Television Comes Along 57
Home Reproduction 59
0.1 for Music? 59
The Bottom Line 63
Calibrating the Monitor System: Frequency Response 64
A Choice of Standardized Response 65
Calibrating the Monitor System: Level 67

3 Multichannel Microphone Technique 71
Introduction 72
Pan Pot Stereo 76
Spaced Omnis 78
Coincident and Near-Coincident Techniques 80
Crossed Figure-8 80
M-S Stereo 82
X-Y Stereo 83
Near-Coincident Technique 83
Binaural 85
Spot Miking 86
Multichannel Perspective 87
Use of the Standard Techniques in Multichannel 88
Surround Technique 91
Surround Microphone Technique for the Direct/Ambient Approach 91
Surround Microphone Technique for the Direct Sound Approach 93
Special Microphones Arrays for 5.1-Channel Recordings 93
Combinations of Methods 97
Some Surround Microphone Setups 100
Simultaneous 2- and 5-Channel Recording 100
Upmixing Stereo to Surround 103
Dynamic Range: Pads and Calculations 103
Virtual Microphones 105

4 Multichannel Mixing and Studio Operations 107
Introduction 108
Mechanics 110

Panners 110
Work Arounds for Panning with 2-Channel Oriented Equipment 112
Panning Law 113
The Art of Panning 115
Non-Standard Panning 117
Panning in Live Presentations 117
A Major Panning Error 118
Increasing the "Size" of a Source 119
Equalizing Multichannel 120
Routing Multichannel in the Console and Studio 121
Track Layout of Masters 122
Double-System Audio with Accompanying Video 123
Reference Level for Multichannel Program 123
Fitting Multichannel Audio onto Digital Video Recorders 124
Multichannel Monitoring Electronics 125
Multichannel Outboard Gear 125
Inter-track Synchronization 127
Requirements for Equipment and Monitor Systems 129
Program Monitoring 131
Postproduction Formats 132
Track Layout 133
Postproduction Delivery Formats 134
Surround Mixing Experience 135
One Case Study: Herbie Hancock's "Butterfly" in 10.2 137
Surround Mixing for DVD Music Videos 139
George Massenburg 140
Multi-Grammy Winner, Music Producer & Engineer, and Equipment and Studio Design Engineer 140

5 Delivery Formats 141
Introduction 143
New Terminology 144
Audio Coding 145
Cascading Coders 148
Sample Rate and Word Length 149
Metadata 151
Multiple Streams 152
Three Level-Setting Mechanisms 154
Dialnorm 154
Dynamic Range Compression 158

Night Listening 159
Mixlevel 159
Audio Production Information Exists 160
Room Type 160
Dolby Surround Mode Switch 161
Downmix Options 161
Level Adjustment of Film Mixes 162
Lip-Sync and Other Sync Problems 163
Reel Edits or Joins 164
Media Specifics 164
Digital Versatile Disc 165
Audio on DVD-Video 167
HD DVD and Blu-Ray Discs 169
Digital Terrestrial and Satellite Broadcast 169
Downloadable Internet Connections 171
Video Games 171
Digital Cinema 172

6 Psychoacoustics 177
Introduction 178
Principal Localization Mechanisms 178
The Minimum Audible Angle 180
Bass Management and Low-Frequency Enhancement Pyschoacoustics 180
Effects of the Localization Mechanisms on 5.1-Channel Sound 182
The Law of the First Wavefront 184
Phantom Image Stereo 184
Phantom Imaging in Quad 185
Localization, Spaciousness, and Envelopment 187
Lessons from Concert Hall Acoustics 188
Rendering 5 Channels Over 2: Mixdown 188
Auralization and Auditory Virtual Reality 190
Beyond 5.1 191

Addendum: The Use of Surrounds in *Saving Private Ryan* 195
Overcoming the Masking Effect 196
Orientation 196
Contrast 197
Movement of Sounds 197
The Limitations of Surrounds 198

Appendix 1: Sample Rate 201
Conclusion 210
What's Aliasing? 211
Definitions 212
MultiBit and One-Bit Conversion 212
Converter Tests 213

Appendix 2: Word Length, Also Known as Bit Depth or Resolution 215
Conversion 215
Dither to the Rescue 216
Dynamic Range 218
Actual Performance 220
How Much Performance Is Needed? 221
Oversampling and Noise Shaping 222
The Bottom Line 223
Analog Reference Levels Related to Digital Recording 224

Appendix 3: Music Mostly Formats 227
Digital Theater Systems CD 227
DVD-Audio 227
Super Audio CD 232
Intellectual Property Protection 232
Towards the Future 233

Index 235

Preface to the Second Edition

It has been 8 years since the first edition of this book was published. In that time, surround sound has grown enormously overall, but with setbacks for some areas such as surround music since, among other things, the lifestyle that accompanies sitting down and listening to music as a recreation has given way to music on the fly.

In those intervening years, a lot of microphone techniques have been developed that were not in the first edition. Mixing has not been as affected perhaps, but the number of places where it is practiced has grown enormously, and the need to accommodate to older consoles has been reduced as those have been replaced by surround capable ones. Delivery formats have grown and shrunk too, as the marketplace decides on what formats it is going to support. At this writing, HD-DVD and Blu-ray are about 1 year old in the marketplace, and it is not clear whether they will become prominent, or possibly give way to legal Internet downloading of movies, which is just beginning. Nevertheless all newer formats generally support at least the capability of the well-established channel layout 5.1. It is interesting to see the media rush to deliver multichannel formats to the home: over-the-air digital television, HD cable, HD satellite, packaged media from DVD-V through the newer formats, and direct fibre to the home all can have a 5.1-channel soundtrack, so surround sound is with us to stay. Recognizing the current state of sales of surround music, I have retained the information on those formats but put them into an appendix. Chapter 1 explains why I think we may not have heard the final word on surround music, since surround itself has risen from the ashes before, and is now widely successful overall.

Two extremely good surround practitioners have added content for this book, and I am indebted to them. George Massenburg was interviewed in his remarkable studio in Nashville and his interview is available as a web-based addition to this book, and Gary Rydstrom gave me his article on surround in *Saving Private Ryan*, an extraordinary view into his thinking about the surround art, which is the Addendum. Both men are at the top of their fields.

Surround has grown beyond the capability of one person to be expert in all the areas. Thus I have tried to vet the various chapters with help from specific experts in their fields, and to consider and reflect their expert opinions. However, the final text is mine. Those particularly helpful to me, in the order of what they did in this book, were Floyd Toole, Florian Camerer, Bob Ludwig, Lorr Kramer, Roger Dressler, and Stanley Lipshitz. My colleague at USC Martin Krieger read the text so that I could understand how it would be understood from someone outside the professional audio field, and he provided useful insights.

As always, I am indebted to my life partner Friederich Koenig, who alternately drives me to do it, and finds that I spend too much time in front of the computer.

1 Introduction

Recorded multichannel sound has a history that dates back to the 1930s, but in the intervening years up to the present it became prominent, died out, became prominent again, and died out again, in a cycle that's had at least four peaks and dips. In fact, I give a talk called "The History and Future of Surround Sound" with a subtitle "A Story of Death and Resurrection in Five Acts" with my tongue barely in my cheek. In modern times we see it firmly established in applications accompanying a picture for movies and television, but in other areas such as purely for music reproduction broad success has been more elusive. However using history as a guide to the future there may well be a broader place for surround music in coming years.

The purpose of this book is to inform recording engineers, producers, and others interested in the topic, about the specifics of multichannel audio. While many books exist about recording, post production, etc., few yet cover multichannel and the issues it raises in any detail. Good practice from conventional stereo carries over to a multichannel environment; other practices must be revised to account for the differences between stereo and multichannel. Thus, this book does not cover many issues that are to be found elsewhere and that have few differences from stereo practice. It does consider those topics that differ from stereo practice, such as how room acoustics have to be different for multichannel monitoring, for instance.

Five-point-one channel sound is the standard for multichannel sound in the mass market today. The 5 channels are left, center, right, left surround, and right surround. The 0.1 channel is called Low-Frequency Enhancement, a single low-frequency only channel with 10 dB greater headroom than the five main channels. However, increasing pressure exists on the number of audio channels, since spatial "gamut"[1] (to borrow a term from our video colleagues) is easily audible to virtually everyone. Seven-point-one channel sound is in fairly widespread use, and will be defined later. My colleagues and I have been at work on "next generation" 10.2-channel sound, for some years, and it also will be treated later.

A Brief History

The use of spatial separation as a feature of composition probably started with the call-and-response form of antiphonal music in the medieval period, an outgrowth of Gregorian chant. Around 1550, Flemish composer Adrian Willaert working in Venice used a chorus in left and right parts for antiphonal singing, matching the two organ chambers oriented to either side of the altar at St. Mark's Basilica. More complex spaced-antiphonal singing grew out of Willaert's work by Giovanni Gabrieli beginning about 1585, when he became principal organist at the same church. He is credited with being the first to use precise directions for musicians and their placement in more than simple left-right orientation, and this was highly suited to the site, with its cross shape within a nearly square footprint (220 × 250′). In the developing polyphonic style, the melodic lines were kept more distinct for listeners by spatially separating them into groups of musicians. The architecture of other churches was affected by the aural use of space as were those in Venice, including especially Freiburg Cathedral in Germany, finished in 1513, which has four organ cabinets around the space. Medieval churches at Chartres, Freiburg, Metz, and Strasbourg had "swallow's nest" organs, located high up in the nave (the tallest part of the church) "striving to lift music, embodied in the organ, high up into the light-filled interior, as a metaphor for musica divina".[2]

The Berlioz *Symphonie Fantastique* (1830) contains the instruction that an oboe is to be placed off stage to imply distance. The composer's *Requiem* (1837) in the section "Tuba mirum" uses four small brass-wind orchestras called "Orchestra No. I to the North, Orchestra No. II to the East, Orchestra No. III to the West, and Orchestra No. IV to the South," emphasizing space, beginning the problem of the assignment of directions to channels! The orchestration for each of the four orchestras is different.

Off-stage brass usually played from the balcony is a feature of Gustav Mahler's Second Symphony *Resurrection*, and the score calls for some of the instruments "in the distance". So the idea of surround sound dates

[1]Meaning the range of available colors to be reproduced; here extended to mean the range of available space.
[2]http://www.koelner-dommusik.de/index.php?id=7&L=6

back at least half a millennia, and it has been available to composers for a very long time.

The original idea for stereo reproduction from the U.S. perspective[3] was introduced in a demonstration with audio transmitted live by high-bandwidth telephone lines from the Academy of Music in Philadelphia to the then four-year old Constitution Hall in Washington, DC on April 27, 1933. For the demonstration, Bell Labs engineers described a 3-channel stereophonic system, including its psychoacoustics, and wavefront reconstruction that is at the heart of some of the "newer" developments in multichannel today. They concluded that while an infinite number of front loudspeaker channels was desirable, left, center, and right loudspeakers were a "practical" approach to representing an infinite number. There were no explicit surround loudspeakers, but there was the fact that reproduction was in a large space with its own reverberation, thus providing enveloping sound acoustically in the listening environment.

Leopold Stokowski was the music director of the Philadelphia Orchestra at the time, but it is interesting to note that he was not in Philadelphia conducting, but rather in Washington, operating 3-channel level and tone controls to his satisfaction. The loudspeakers were concealed from view behind an acoustically transparent but visually opaque scrim-cloth curtain. A singer walked around the stage in Philadelphia, and the stereo transmission matched the location. A trumpeter played in Philadelphia antiphonally with one on the opposite side of the stage in Washington. Finally the orchestra played. The curtain was raised, and to widespread amazement the audience in Washington found they were listening to loudspeakers!

Stokowski was what we would today call a "crossover" conductor, interested in bringing classical music to the masses. In 1938 he was having an affair with Greta Garbo, and to keep it reasonably private, wooed her at Chasen's restaurant in Hollywood, a place where discretion ruled. Walt Disney, having heard "The Sorcerer's Apprentice" in a Hollywood Bowl performance and fallen for it had already had the idea for *Fantasia*. According to legend, one night at Chasen's Disney prevailed upon a waiter to deliver a note to Stokowski saying he'd like to meet. They turned out already to be fans of each other's work, and Disney pitched Stokowski on his idea for a classical music animated film. Stokowski was so enthusiastic that he recorded the music for *Fantasia* without a fee, but he also told Walt he wanted to do it in stereo, reporting to him on the 1933 experiment. Walt Disney, sitting in his

[3]Alan Blumlein, working at EMI in England in the same period also developed many stereo techniques, including the "Blumlein" pair of crossed figure-8 microphones. However, his work was in two channel stereo.

living room in Toluca Lake some time later thought that during "The Flight of the Bumblebee," not only should the bumblebee be localizable across the screen, but also around the auditorium. Thus surround sound was born. Interestingly "The Flight of the Bumblebee" does not appear in *Fantasia*. It was set aside for later inclusion in a work that Walt saw as constantly being revised and running for a long, long time. So even that which ends up on the proverbial cutting room floor can be influential, as it here invented an industry. The first system of eight different variations that engineers tried used three front channels located on the screen, and two surround channels located in the back corners of the theater—essentially the 5.1-channel system we have today. During this development, Disney engineers invented multitrack recording, pan potting, overdubbing, and when it went into theaters, surround arrays. Called Fantasound, it was the forerunner of all surround sound systems today.

While guitarist Les Paul is usually given the credit for inventing overdubbing in the 1950s, what he really invented was Sel-Sync recording, that is, playing back off the record head of some channels of a multitrack tape machine, used to cue (by way of headphones) musicians who were then recorded to other tracks; this kept the new recordings in sync with the existing ones. In fact, overdub recording by playing back an existing recording for cueing purposes over headphones was done for *Fantasia*, although it was before tape recording was available in the U.S. Disney used optical sound tracks, which had to be recorded, then developed, then printed and the print developed, and finally played back to musicians over headphones who recorded new solo tracks that would run in sync with the original orchestral recording. Then, after developing the solo tracks, and printing and developing the prints, both the orchestral and solo tracks were played simultaneously on synchronized optical dubbers (also called "dummies") and remixed. By changing the relative proportions of orchestra and solo tracks in the mix the ability to vary the perspective of the recording from up-front solos to full orchestra was accomplished.

So multichannel recording has a long history of pushing the limits of the envelope and causing new inventions, which continues to the present day. By the way, note that multichannel is different from multitrack. Multichannel refers to media and sound systems that carry multiple loudspeaker channels beyond two. Multitrack is, of course, a term applied to tape machines that carry many separate tracks, which may, or may not, be organized in a multichannel layout.

Fantasia proved to be a one-shot trial for multichannel recording before World War II intruded. It was not financially successful in its first run, and the war intervened too soon for it to get a toehold in the marketplace. The war resulted in new technology introductions that proved to be useful in the years following the war. High-quality permanent magnets developed for airplane crew headphones made better theater loudspeakers possible (loudspeakers of the 1930s had to have DC supplied to them to make a magnetic field). Magnetic recording on tape and through extension magnetic film was war booty, appropriated from the Germans. Post-war a pre-war invention was to come on the scene explosively, laying wreck to the studio system that was vertically integrated from script writing through theatrical exhibition and that had dominated entertainment in the 1930s—television. Between 1946 and 1954 the number of patrons coming to movie theaters annually fell by about one-half, largely due to the influence of television.

Looking for a post-war market for film-based gunner training simulators, Fred Waller produced a system shot on three parallel strips of film for widescreen presentation and having 7 channels of sound developed by Hazard Reeves called Cinerama.[4] Opening in 1952 with *This is Cinerama*, it proved a sensation in the few theaters that could be equipped to show it. The 7 channels were deployed as five across the screen, and two devoted to surrounds. Interestingly, during showing a playback operator could switch the surround arrays from the two available source tracks to left and right surrounds, or front sides and back surrounds, thus anticipating one future development by nearly 50 years.

In 1953 20th Century Fox, noting both the success of Cinerama but also its uneconomic nature, and finding the market for conventional movies shrinking, combined the 1920s anamorphic photography invention by French Professor Henri Chrétien, with four-track magnetic recordings striped along the edges of the 35 mm release print, to produce a wide screen system that could have broader release than Cinerama films. They called it Cinemascope. The 4-channel system called for three screen channels, and one channel directed to loudspeakers in the auditorium. Excess noise in the magnetic reproduction of the time was partially overcome by the use of high-frequency standardized playback rolloff called generically the Academy curve. When this proved to be not enough filtering so that hiss was still heard from the "effects" loudspeakers during quiet passages, a high-frequency 12 kHz sine-wave switching tone was recorded on the track in addition to the audio to trigger the effects channel on and off. Played with a notch filter for the audio so that the tone would not be audible, simultaneously a controlling parallel side chain was peaked up at the tone frequency and

[4] A good documentary about the subject is *Cinerama Adventure*, http://www.cineramaadventure.com/

used to decide when to switch these speakers on and off. Today these speakers form what we call the surround loudspeakers.

In 1955, a six-track format on 70 mm film was developed by the father and son team Michael Todd and Michael Todd, Jr. for road show presentations of movies. Widescreen photography was accomplished for Todd AO[5] without the squeezing of the image present in Cinemascope through the simple expedient of making the film twice as wide. The six tracks were deployed as five screen channels as in Cinerama, and one surround channel. Interestingly, both the Todds had worked at Cinerama producing the most famous sequence in *This is Cinerama*, the roller coaster ride. They realized that the three projectors and one sound dubber and all the operators needed by Cinerama could be replaced by one wide format film with six magnetic recordings on four stripes, with two tracks outside the perforations and one inside, on each side of the film, thus lowering costs of exhibition greatly while maintaining very high performance. The idea was described as "Cinerama from one hole." In fact the picture quality of the 1962 release of *Lawrence of Arabia* shot in 65 mm negative for 70 mm release has probably never been bested.

While occasional event road shows like *2001* (1968) continued to be shot in 65 mm, the high cost of striping release prints put a strong damper on multichannel recording for film, and home developments dominated for years from the mid-1950s through the mid-1970s. Stereo came to the home as a simplification of theater practice. While three front channels were used by film-based stereophonic sound, a phonograph record only has two groove walls that can carry, without complication, two signals. Thus, two channel stereo reproduction in the home came with the introduction of the stereo LP in 1958, and other formats followed because there were only two loudspeakers at home. FM radio, various tape formats, and the CD followed the lead established by the LP.

Two-channel stereo relied upon the center of the front sound stage to be reproduced as what is called a phantom image. These sounds made a properly set-up stereo seem almost magical, as a sound image floated between the loudspeakers. Unfortunately, this phantom effect is highly sweet-spot dependent, as the listener must sit along the centerline of the speakers, and the speakers and room acoustics for the two sides must be well matched. It was good for a single listener experience, and drove a hobbyist market for a long time, but was not very suitable for multiple listeners. The reasons for this are described in Chapter 6 on Psychoacoustics.

[5]The Todds partnered with the American Optical company for Todd AO.

The Quad era of the late 1960s through 1970s attempted to deliver four signals through the medium of two tracks on the LP using either matrices based on the amplitude and phase relations between the channels, or on an ultrasonic carrier requiring bandwidth off the disc of up to 50 kHz. The best known reasons for the failure of Quad include the fact that there were three competing formats on LP, and one tape format, and the resistance on the part of the public to put more loudspeakers into their listening rooms. Less well known is the fact that Quad record producers had very different outlooks as to how the medium should be used, from coming closer to the sound of a concert hall than stereo by putting the orchestra up front and the hall sound around the listener, to placing the listener "inside the band," a new perspective for many people that saw both tasteful, and some not so tasteful, presentations. Also, the psychoacoustics of a square array of loudspeakers were not well understood at the time, leading to the absurdity that some people still think: that Quad is merely stereo reproduced in four quadrants, so that if you face each way of a compass in turn you have a "stereo" sound field. The problem with this though is that we can't face the four quadrants simultaneously. Side imaging works much differently from front and back imaging because our ears are at the two sides of our heads. The square array would work well if we were equipped by nature with four ears at 90° intervals about our heads! When it came to be studied by P.A. Ratliff at the BBC,[6] it is fair to say that the quadraphonic square array got its comeuppance.

The misconception that Quad works as four stereo systems ranks right up there with the inanity that since we have two ears, we must need only two loudspeakers to reproduce sound from every direction—after all, we hear all around don't we? This idea neglects the fact that each loudspeaker channel does not have input to just one ear, but two, through "crosstalk" components consisting of the left loudspeaker sound field reaching the right ear, and vice versa. Even with sophisticated crosstalk cancellation, which is difficult to make anything other than supremely sweet-spot sensitive, it is still hard to place sounds everywhere around a listener. This is because human perception of sound uses a variety of cues to assign localization to a source, including interaural level differences, that is, differences in level between the two ears; interaural time differences; the frequency response for each angle of arrival associated with a sound field interacting with the head, outer ears called the pinnae, shoulder bounce, etc. and collectively called the head-related transfer function (HRTF); and dynamic variations on these static cues since our heads are rarely clamped down

[6] http://www.bbc.co.uk/rd/pubs/reports/1974-38.pdf

to a fixed location. The complexity of these multiple cue mechanisms confounds simple-minded conclusions about how to do all-round imaging from two loudspeakers.

While home sound developed along a two channel path, cinema sound enjoyed a revival starting in the middle 1970s with several simultaneous developments. The first of these was the enabling technology to record 4 channels worth of information on an optical sound track that had space for only two tracks and maintain reasonable quality. Avoiding the expensive and time-consuming magnetic striping step in the manufacture of prints, and having the ability to be printed at high speed along with the picture, stereo optical sound on film made other improvements possible. For this development, a version of an amplitude-phase matrix derived from one of the quadraphonic systems but updated to film use was used. Peter Scheiber's amplitude-phase matrix method was the one selected. Dubbed Dolby Stereo in this incarnation, it was a fundamental improvement to optical film sound, combining wider frequency and dynamic ranges through the use of companding noise reduction with the ability to deliver two tracks containing 4 channels worth of content by way of the matrix. In today's world of some 30 years later, the resulting dynamic range and difficulties with recording for the matrix make this format seem dated, but at the time, coming out of the mono optical sound on film era, it was a great step forward.

The technical developments would probably not have been sustained—after all, it cost more—if not for particularly good new uses that were found immediately for the medium. *Star Wars*, with its revolutionary sound track, followed in six months by *Close Encounters of the Third Kind*, cemented the technical developments in place and started a progression of steady improvements. At first Dolby A noise reduction was used principally to gain bandwidth. Why a noise reduction system may be used to improve bandwidth may not be immediately obvious. Monaural sound tracks employed a great deal of high-frequency rolloff in the form of the Academy filter to overcome the hiss associated with film grain running past the optical reader head. Simply extending the bandwidth of such tracks would reveal a great deal of hiss. By using Dolby A noise reduction, particularly the 15 dB it achieved at high frequencies, bandwidth could be extended from effectively about 4 to 12 kHz while keeping the noise floor reasonable and bringing film sound into the modern era. Second generation companding signal processing, Dolby SR, was subsequently used on analog sound tracks starting in 1986, permitting an extension to 16 kHz, greater headroom versus frequency, lower distortion, and noise performance such that the sound track originated hiss remained well below the noise floor of good theaters.

With the technical improvements to widespread 35 mm release prints came the revitalization of the older specialty format, 70 mm. Premium prints playing in special theaters offered a better experience than the day-to-day 35 mm ones. Using six-track striped magnetic release prints, running at 22.5 ips, with wide and thick tracks, the medium was unsurpassed for years in delivery of wide frequency and dynamic ranges, and multichannel, to the public at large. In the 1970s Dolby A noise reduction allowed a bandwidth extension in playback of magnetic tracks, just as it had with optical ones, and some use was made of Dolby SR on 70 mm when it became available.

For 70 mm prints of *Star Wars*, the producer Gary Kurtz realized along with Dolby personnel including Steve Katz, that the low-frequency headroom of many existing theater sound systems was inadequate for what was supposed to be a war (in space!). Reconfiguring the tracks from the original six-track Todd AO 70 mm format, engineers set a new standard: three screen channels, one surround channel, and a "Baby Boom" channel, containing low-frequency only content, with greater headroom accomplished by using separate channels and level adjustments. An extra low-frequency only channel with greater headroom proved to be a better match to human perception. This is because hearing requires more energy at low frequencies to sound equally as loud as the mid-range. *Close Encounters of the Third Kind* then used the first dedicated subwoofers in theaters (*Star Wars* used the left-center and right-center loudspeakers from the Todd AO days that were left over in larger theaters), and the system became standard.

Two years after these introductions, *Superman* was the first film since Cinerama to split the surround array in theaters into two, left and right. The left/right high-frequency sound for the surrounds was recorded on the left-center and right-center tracks, while the boom track continued to be recorded on these tracks at low frequencies. Filters were used in cinema processors to split bass off from treble to produce a boom channel and stereo surrounds. For theaters equipped only for mono surround, the wide range surround track on the print sufficed. So one 70 mm stereo-surround print was compatible for playback in both mono- and stereo-surround equipped theaters. In the same year, *Apocalypse Now* made especially good artistic use of stereo surround, and most of the relatively few subsequent 70 mm releases used the format of three screen channels, two surround channels, and a boom channel.

In 1987, when a subcommittee of the Society of Motion Picture and Television Engineers (SMPTE) looked at putting digital sound on film, meetings were held about the requirements for the system. In a series of meetings and documents, the 5.1-channel system emerged as being the minimum number of channels that would create the sensations desired from a new system, and the name 5.1 took hold from that time. In fact, this can be seen as a codification of existing 70 mm practice that already had five main channels and one low-frequency only, higher headroom channel.

With greater recognition among the public of the quality of film sound that was started by *Star Wars*, home theater began, rather inauspiciously at first, with the coming of two channel stereo tracks to VHS tape and Laser Disc. Although clearly limited in dynamic range and with other problems, early stereo media were quickly exploited as carriers for two-channel encoded Dolby Stereo sound tracks originally made for 35 mm optical film masters, called Lt/Rt (left total, right total, used to distinguish encoded tracks from conventional stereo left/right tracks), in part because the Lt/Rt was the only existing two channel version of a movie, and copying it was the simplest thing to do to make the transfer from film to video.

Both VHS and Laser Disc sound tracks were enhanced when parallel, and better quality, two channel stereo recording methods were added to the media. VHS got its "Hi-Fi" tracks, recorded in the same area as the video and by separate heads on the video scanning drum as opposed to the initial longitudinal method, which suffered greatly from the low tape speed, narrow tracks, and thin oxide coating needed for video. The higher tape-to-head speed, FM recording, and companding noise reduction of the Hi-Fi tracks contributed to the more than 80 dB signal-to-noise ratio, a big improvement on the "linear" tracks, although head and track mismatching problems from recorder to player could cause video to "leak" into the audio and create an annoying variable buzz, modulated by the audio. Also, unused space was found in the frequency spectrum of the signals on the Laser Disc to put in a pair of 44.1 kHz, 16-bit linear PCM tracks, offering the first medium to deliver digital sound in the home accompanying a picture.

The two higher-quality channels carried the Dolby Stereo encoded sound tracks of more than 10,000 films within a few years, and the number of home decoders, called Dolby Pro Logic, is in the many tens of millions. This success changed the face of consumer electronics in favor of the multichannel approach. Center loudspeakers, many of dubious quality at first but also offered at high quality, became commonplace, as did a pair of surround loudspeakers added to the left-right stereo that many people already had.

Thus, today there is already a playback "platform" in the home for multichannel formats. The matrix, having served long and well and still growing, nevertheless is probably nearing the end of its technological lifetime.

Problems in mixing for the matrix include:

- Since the decoder relies on amplitude and phase differences between the 2 channels, interchannel amplitude or phase difference arising from errors in the signal path leads to changes in spatial reproduction: a level imbalance between Lt and Rt will "tilt" the sound towards the higher channel, while phase errors will usually result in more content coming from the surrounds than intended. This problem can be found at any stage of production by monitoring through a decoder.
- The matrix is very good at decoding when there is one principal direction to decode at one time, but less good as things get more complex. One worst case is separate talkers originating in each of the channels, which cause funny steering artifacts to happen (parts of the speeches will come from the wrong place).
- Centered speech can seem to modulate the width of a music cue behind the speech. An early scene in *Young Sherlock Holmes* is one in which young Holmes and Watson cross a courtyard accompanied by music. On a poor decoder the width of the music seems to vary due to the speech.
- The matrix permits only 4 channels. With three used on the screen, that leaves only a monaural channel to produce surround sound, a contradiction. Some home decoders use means to decorrelate the mono-surround channel into two to overcome the tendency of mono surround to localize to the closer loudspeaker, or if seated perfectly centered between matched loudspeakers, in the center of your head.

Due to these problems, professionals in the film industry thought that a discrete multichannel system was desirable compared to the matrix system.

The 1987 SMPTE subcommittee direction towards a 5.1-channel discrete digital audio system led some years later to the introduction of multiple digital formats for sound both on and off release prints. Three systems remain after some initial shakeout in the industry: Dolby Digital, Digital Theater Systems (DTS), and Sony Dynamic Digital Sound (SDDS). Dolby Digital provided 5.1 channels in its original form, while SDDS has the capacity for 7.1 channels (adding two intermediate screen channels, left center, and right center). While DTS units could be delivered initially with up to 8 channels as an option, most of the installed base is 5.1. The relatively long gestation period for these systems was caused by, among other things, a fact of life: there was not enough space on either the release prints or on double-system CD-ROM followers to use conventional linear PCM coding. Each of the three systems uses one method or another of low-bit-rate coding, that is, of reducing the number of bits that must be stored compared to linear PCM.

The 5.1 channels of 48 kHz sampled data with 18-bit linear PCM coding (the recommendation of the SMPTE to accommodate the dynamic range necessary in theaters, determined from good theater background floor measurements to the maximum undistorted level desired, 105 dB SPL per channel) requires:

$$\begin{array}{l}5.005 \text{ channels} \times 48\,\text{k samples/s} \\ \text{for 1 channel} \times 18 \text{ bits/sample}\end{array} = 4{,}324{,}320 \text{ bits/s.}$$

(The number 5.005 is correct; 5.1 was a number chosen to represent the requirement more simply. Actually, 0.005 of a channel represents a low-frequency only channel with a sample rate of 1/200 of the principal sample rate.)

In comparison, the compact disc has an audio payload data rate of 1,411,200 bits/s. Of course, error coding, channel coding, and other overhead must be added to the audio data rate to determine the entire rate needed, but the overhead rate is probably a similar fraction for various media.

Contrast the 4.3 million bits per second needed to the data rate that can be achieved due to the space on the film. In Dolby Digital, a data block of 78 bits by 78 bits is recorded between each perforation along one side of the film. There are 4 perforations per frame and 24 frames per second yielding 96 perforations per second. Multiplying 78 × 78 × 96 gives a data capacity off film of 584,064 bits/s, only about 1/7 that needed just for the audio, not to mention error correcting overhead, synchronization bits, etc. While other means could be used to increase the data rate, such as shrinking the bit size, using other parts of the film, or use of a double system with the sound samples on another medium, Dolby Labs engineers chose to work with bits of this size and position for practical reasons. The actual payload data rate they used on film is 320,000 bits/s, 1/13.5 of the representation in linear PCM. Bit-rate reduction systems thus became required, in this case because the space on the film was so limited. Other media had similar requirements, and so do broadcast channels and point-to-point communications such as the Internet.

While there are many methods of bit-rate reduction, as a general matter, those having a smaller amount of reduction use waveform-based methods, while those using larger amounts of reduction employ psychoacoustics, called perceptual coding. Perceptual coding makes use of the fact that louder sounds cover up softer ones, especially those close by in frequency, called frequency masking. Loud sounds not only affect softer sounds presented simultaneously, but also mask those sounds that precede or follow them through temporal masking. By dividing the audio into frequency bands, then processing each band separately, just coding each of the bands with the number of bits necessary to account for masking in frequency and time domains, the bit rate is reduced. At the other end of the chain, a symmetrical digital process reconstructs the audio in a manner that may be indistinguishable from the original, even though the bit rate has been reduced by a factor of more than 10.

The transparency of all lower than PCM bit-rate systems is subject to potential criticism, since they are by design losing data.[7] The only unassailable methods that reveal differences between each of them and an original recording, and among them, are complex listening tests based on knowledge of how the coders work. Experts must select program material to exercise the mechanisms likely to reveal sonic differences, among other things.

There have been a number of such tests, and the findings of them include:

- The small number of multichannel recordings available gave a small universe from which to find programs to exercise the potential problems of the coders—thus custom test recordings are necessary.
- Some of the tradeoffs involved in choosing the best coder for a given application include audio quality, complexity and resultant cost, and time delay through the coding process.
- Some of the coders on some of the items of program material chosen to be particularly difficult to code are audibly transparent.
- None of the coders tested is completely transparent all of the time, although the percentage of time in a given channel carrying program material that will show differences from the original is unknown. For instance, one of the most sensitive pieces of program material is a simple pitch pipe, because its relatively simple spectrum shows up quantizing noise arising from a lack of bits available to assign to frequencies "in between" the harmonics, but how much time in a channel is devoted to such a simple signal?
- In one of the most comprehensive tests, film program material from *Indiana Jones and the Last Crusade*, added after selection

[7]Of course no data is deliberately lost, but rather lossy coders rely on certain redundancies in the signal and/or human hearing perceptual mechanisms, to reduce the bit rate. Such coders are called "lossy" because they cannot reconstruct the original data.

> of the other particularly difficult program material, showed differences from the original. This was important because it was not selected to be particularly difficult, and yet was not the least sensitive among the items selected. Thus experts involved in selection should use not only specially selected material, but also "average" material for the channel.

Bit-rate reduction coders enabled the development of digital sound for motion pictures, using both sound-on-film and sound-on-follower systems. During the period of development of these systems, another medium was to emerge that could make use of the same coding methods. High-definition television made the conceptual change from analog to digital when one system proponent startled the competition with an all-digital system. Although at the time this looked nearly impossible, a working system was running within short order, and the other proponents abandoned analog methods within a short period of time. With an all-digital system, at first two channels of audio were contemplated. When it was determined that multichannel coded audio could operate at a lower bit rate than two channels of conventional digital audio, the advantages of multichannel audio outweighed concerns for audio quality if listening tests proved adequate transparency for the proposed coders. Several rounds of listening tests resulted in first selection, and then extensive testing, of Dolby AC-3 as the coding method. With the selection of AC-3 for what came to be known as Digital Television, Dolby Labs had a head start in packaged media, first Laser Disc, then DVD-Video. It was later dubbed simply Dolby Digital in the marketplace, rather than relying on the development number of the process, AC-3. DTS also provided alternate coding methods on Laser Disc, DVD-Video, and DTS CD for music.

With the large scale success of DVD, introduced in 1997, and with music CDs flagging in sales due to, among other things, lack of technical innovation, two audio-mostly packaged media were introduced: DVD-A in 2001 and Super Audio Compact Disc (SACD) in 2003. Employing different coding methods, insofar as surround sound operates the two are pretty similar, delivering up to six wide range channels, usually employed as carriers for either 2.0- or 5.1-channel audio. Unfortunately, the combined sales of the two formats only added up to about the sales of LP records in 2005, some 1/700th of the number of CDs. There are probably several reasons for this: one is the shift among early adopters away from packaged media to downloadable media to computers and portable music devices. While this shift does not preclude surround sound distribution to be rendered down to two channel headphones as needed, nonetheless the simplicity of two channel infrastructure has won out, for now. Perhaps this parallels the development of surround for movies, where two-channel infrastructure won for years until a discrete 5.1-channel system became possible to deliver with little additional cost.

Over the period from 2000 to 2006 U.S. sales show a decline in CD unit sales from 942.5 million to 614.9 million. The corresponding dollar value is from $13,214.5 million to $9,162.9 million, not counting the effect of inflation. Some of this loss is replaced by digital downloads of singles, totaling 586 million units in 2006, and albums, totaling 27.6 million units. The total dollar volume though of downloads of $878.0 million, added to the physical media sales and mobile and subscription sales, still only produces a total industry volume at retail of $11,510.2 million, an industry shrinkage of 45.7% in inflation adjusted dollars over the period from 1999 to 2006. Accurate figures on piracy due to at first download sites such as Napster and later peer-to-peer networks are difficult to come by, but needless to say contribute to the decline of sales. One survey by market research firm NPD for the RIAA showed 1.3 billion illegal downloads by college students in 2006, and the RIAA has filed numerous lawsuits against illegal downloaders. The survey also claims that college students represent 26% of the total of P2P downloads, so the total may be on the order of 5.2 billion illegal downloads. College students self-reported to NPD that two-thirds of the music they acquired was obtained illegally.

To put the 2006 size of the music business in perspective, the U.S. theatrical box office for movies was $9,490 million in that year, and DVD sales were $15,650 million. Rental video added another $7,950 million, bringing the total annual spent on movies and recorded television programs[8] in the USA to $33,090 million. So adding the various retail values together the motion picture business, much of which is based on 5.1-channel surround sound, is 2.87 times as large as the music industry in the USA in dollar volume.[9] Illegal downloads of movies also affect the bottom line, but less so than that of the music industry simply because the size of the files for movies is much larger and requires longer download times. The Motion Picture Association of America (MPAA) estimates that U.S. losses in 2005 were $1.3 billion, divided between Internet downloads and DVD copies.

A second reason for the likely temporary failure of surround music-mostly media is that the producers ignored a lesson from consumer

[8]Also counts DVD Music Video discs with sales of about $20 million.

[9]The source for these statistics is the 2006 Year-End Shipment Statistics of the Recording Industry Association of America, the Motion Picture Association of America, and an article on January 5, 2007 by Jennifer Netherby on www.videobusiness.com.

research. The Consumer Electronics Association conducts telephone polling frequently. A Gallup poll for CEA found that about 2/3 of listeners prefer a perspective from the best seat in the house, compared to about 1/3 that want to be "in the band." This is a crucial difference: while musicians, producers, and engineers want to experience new things, and there are very real reasons to want to do so that are hopefully explained later in this book, nonetheless much of the audience is not quite ready for what we want to do experimentally, and must be brought along by stages. What music needs is the effect that the opening of *Star Wars* had on audiences in 1977—many remember the surround effect of the space ship arriving coming in over your head, with the picture and sound working in concert, as their first introduction to surround sound.

Further consumer research in the form of focus groups listening to surround music and responding was conducted by the Consumer Electronics Association. With a person-on-the-street audience, surround music was found to range from satisfying through annoying, especially when discrete instruments were placed in the surround channels. Harkening back to the introduction of *Fantasia* when some people left the theater to complain to management that there was a chorus singing from behind them, insecurity on the part of portions of the audience drive them to dislike the experience.

A third reason, and probably the one most cited, is that two competing formats, like the Betamax versus VHS war, could not be sustained in the marketplace.

Meanwhile the Audio Engineering Society Technical Committee on Multichannel and Binaural Audio Technology wrote an information guideline called *Multichannel Surround Sound Systems and Operations*, and published it in 2001. It is available as a free download at http://www.aes.org/technical/documents/AESTD1001.pdf. The document was written by a subcommittee of the AES committee.[10] Contributions and comments were also made by members of the full committee and other international organizations.

HD-DVD and Blu-ray high-definition video discs are now on the market. These second generation packaged media optical discs offer higher picture quality, so enhancements to sound capability was also on the agenda of the developers. Chapter 5 covers these media.

With 5.1 channels of discrete digital low-bit-rate coded audio standard for film, and occasional use made of 7.1 discrete channels including

[10]The Writing Group was: Francis Rumsey (Chair), David Griesinger, Tomlinson Holman, Mick Sawaguchi, Gerhard Steinke, Günther Theile, and Toshio Wakatuki.

left-center and right-center screen channels, almost inexorably there became a drive towards more channels, due to the sensation that only multidirectional sound provides. In 1999, for the release of the next installment in the Star Wars series, a system called Dolby Surround EX was introduced. Applying a specialized new version of the Dolby Surround matrix to the two surround channels resulted in the separation of the surround system into left, back, and right surround channels from the two that had come before. Later, since a 4:2:4 matrix was used but 1 channel was not employed in Dolby Surround EX, one film, *We Were Soldiers*, was made using 4 channels of surrounds: left, right, back, and overhead. DTS also developed a competitive system with 6.1 discrete channels, called DTS ES.

With 5.1-channel sound firmly established technically by the early 1990s and with its promulgation into various media likely to require some years, the author and his colleagues began to study what 5.1-channel audio did well, and in what areas its capabilities could be improved. The result is a system called 10.2. All sound systems have frequency range and dynamic range capabilities. Multichannel systems offer an extension of these capabilities to the spatial domain. Spatial capabilities are in essence of two kinds: imaging and envelopment. These two properties lie along a dimension from pinpoint imaging on the one hand, to completely diffuse and thus enveloping on the other. For the largest seating area of a sound system, imaging is improved as the number of channels goes up. In fact, if you want everyone in an auditorium to perceive sound from a particular position the best way to do that is to put a loudspeaker with appropriate coverage there. This however is probably inconvenient for all sound sources, so the use of phantom images, lying between loudspeakers, becomes necessary. Phantom images are more fragile, and vary more as one moves around a listening space, but they are nonetheless the basis for calling stereophonic sound "solid," which is the very meaning of stereo.

Digital Cinema is now a reality, with increasing number of screens so equipped in the U.S. Audio for it consists of up to 16 linear PCM coded audio tracks, with a sample rate of 48 kHz and a 24-bit word length. Provision is made in the standards to increase the sample rate to 96 kHz, should that ever become required.

Other systems entered the experimental arena too. NHK fielded an experimental system with 22.2 channels, while Europe's Fraunhofer Institute introduced its Iosono wavefield synthesis system harkening back to Bell Labs infinite line of microphones feeding an infinite line of loudspeakers, with 200 loudspeakers representing infinity in one cinema installation.

All in all, upwards pressure on the number of audio channels is expected to continue into the future. This is because the other two competitors[11] for bits, sample rate representing frequency range, and word length also known as resolution and bit depth representing dynamic range, both are what engineers call saturating functions. With a saturating function, when one has enough there is simply no point in adding to the feature. For instance, if 24-bit audio were actually practical to be realized, setting the noise floor at the threshold of hearing would produce peaks of 140 dB SPL, and the entire hearing audience would revolt. I tried playing a movie with relatively benign maximum levels, *Sea Biscuit*, to a full audience at its original Hollywood dubbing stage level in the midwest, and within 10 minutes more than five people had left the full theater to complain to the management about how loud it was, and this was at a level that could not exceed 105 dB SPL/channel, and almost certainly never reached that level. So there can be said to be no point in increasing the bit depth to 24 from, say 20, as the audience would leave if the capability were to be used.

The number of loudspeaker channels on the other hand is not a saturating function, but rather what engineers call an asymptotic function: the quality rises as the number of channels increases, but with declining value as the number of channels goes up. Many people have experienced this, including the whole marketplace. The conversion from mono to stereo happened because all of the audience could hear the benefit; today stereo to 5.1 is thoroughly established as an improvement. Coming at the question from another point of view, English mathematician Michael Gerzon has estimated that it would take 1 million channels to transmit one space at each point within it into another space, while James Johnston then at AT&T Labs estimated that 10,000 channels would do to get the complex sound field reproduced thoroughly in the vicinity of one listener's head. With actual loudspeakers and interstitial phantom images, Jens Blauert in his seminal book *Spatial Hearing* has said that about 30 channels would do placed on the surface of 1/2 of a sphere in the plane above and around a listener for listening in such a half-space.

The sequence 1, 2, 5, 10 is familiar to anyone who has ever spent time with an oscilloscope: the voltage sensitivity and sweep time/div knobs are stepped in these quantized units. They are approximately equal steps along a logarithmic scale. G.T. Fechner showed in 1860 that just noticeable differences in perception were smaller at lower levels and

---

[11]This is not strictly true because various low-bit-rate coding schemes compared to linear PCM ameliorate the need for bits enough that they make practical putting sound on film optically for instance; but in the long run, it is really bandwidth and dynamic range that are competitors for space on the medium.

larger at bigger ones. He concluded that perception was logarithmic in nature for perceptual processes. Applying Fechner's law today, one would have to say that the sequence of channels 1, 2, 5 10, 20 ... could well represent the history of the number of channels when viewed in the distant future.

A summary of milestones in the history of multichannel surround sound is as follows:

- Invention in 1938 with one subsequent movie release *Fantasia* (3 channels on optical film were steered in reproduction to front and surround loudspeakers).
- Cinerama 7 channel, Cinemascope 4 channel, and Todd AO 6 channel introduced in short order in the 1950s, but declining as theater admissions slipped through the 1960s.
- Introduction of amplitude–phase matrix technology in the late 1960s, with many subsequent improvements.
- Revitalization in 1975–1977 by 4-channel matrixed optical sound on film.
- Introduction of stereo surround for 70 mm prints in 1979.
- Introduction of stereo media for video, capable of carrying Lt/Rt matrixed audio, and continuing improvements to these media, from the early 1980s.
- Specification for digital sound on film codified Five point one channel sound in 1987.
- Standardization on 5.1 channels for Digital Television in the early 1990s.
- Introduction of three competing digital sound for film systems in the early 1990s, see Fig. 1-2 for their representation on film.
- Introduction of 5.1-channel digital audio for packaged media, in the late 1990s.
- Introduction of matrixed 3-channel surround separating side from back surround using two of the discrete channels of the 5.1-channel system with an amplitude–phase matrix in 1999.
- Introduction of 10.2-channel sound in 1999.
- Consumer camcorder recording 5.1-channel audio from built-in microphone, Christmas 2005.
- Blu-ray and HD-DVD formats including multichannel linear PCM among other standards, 2006.
- Digital Cinema, 2006.

Fig. 1-1, a timeline, gives additional information.

In the history of the development of multichannel, these milestones help us predict the future; there is continuing pressure to add channels because more channels are easily perceived by listeners. Anyone

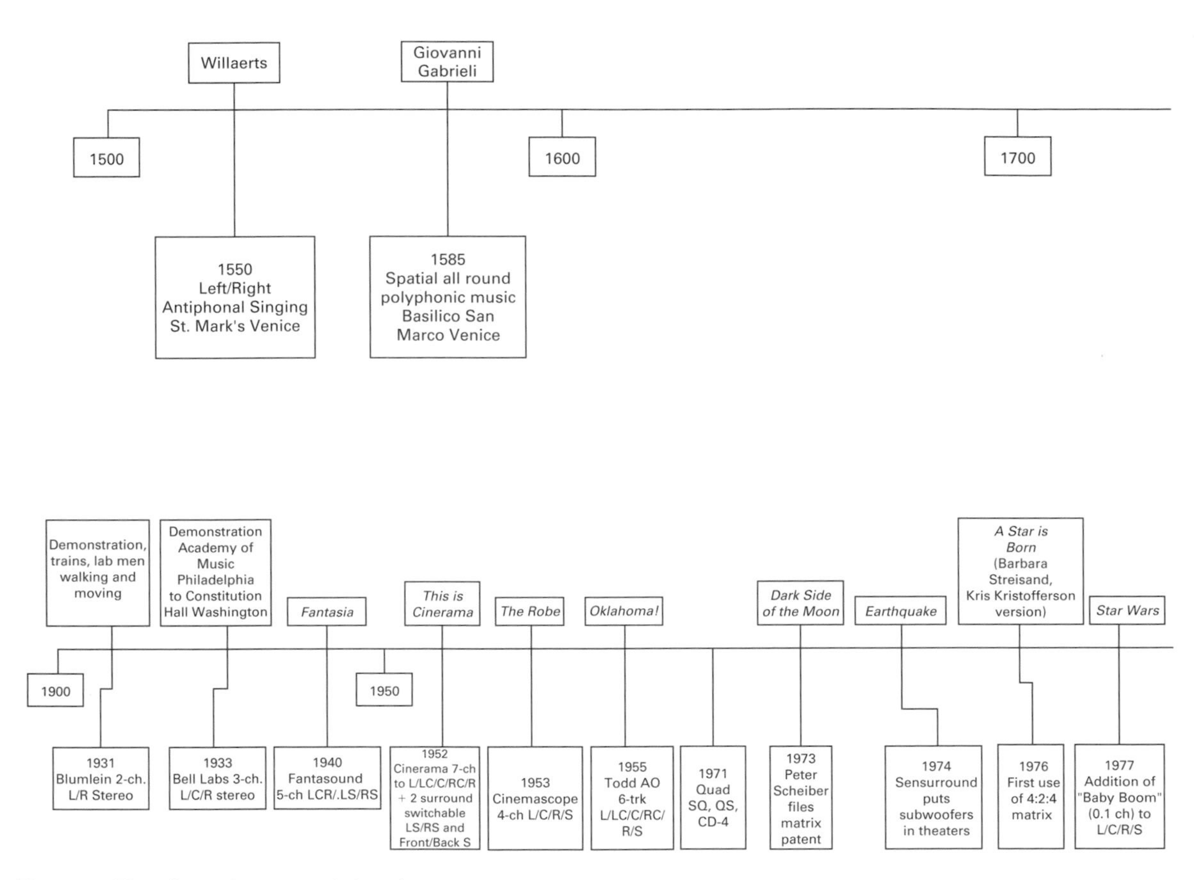

Fig. 1-1 Timeline of surround developments.

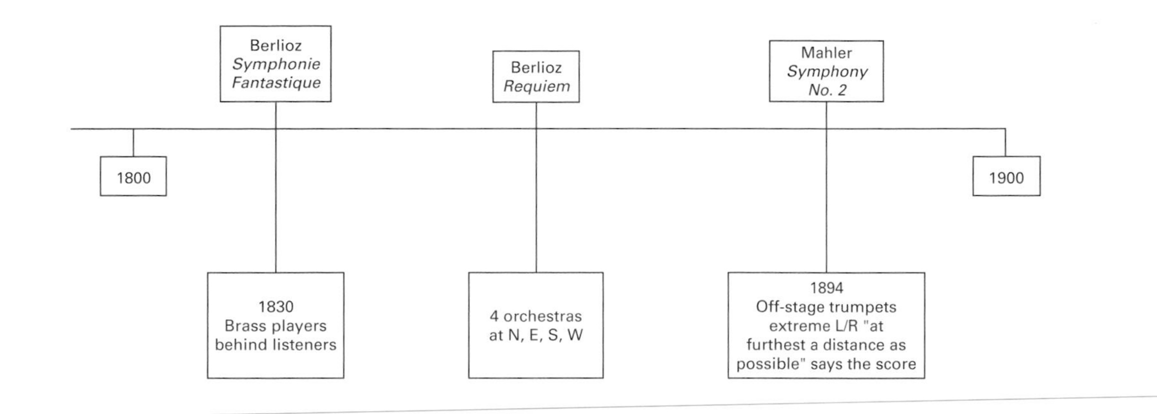
Berlioz
Symphonie Fantastique
Berlioz
Requiem
Mahler
Symphony No. 2
1800
1900
1830
Brass players behind listeners
4 orchestras at N, E, S, W
1894
Off-stage trumpets extreme L/R "at furthest a distance as possible" says the score

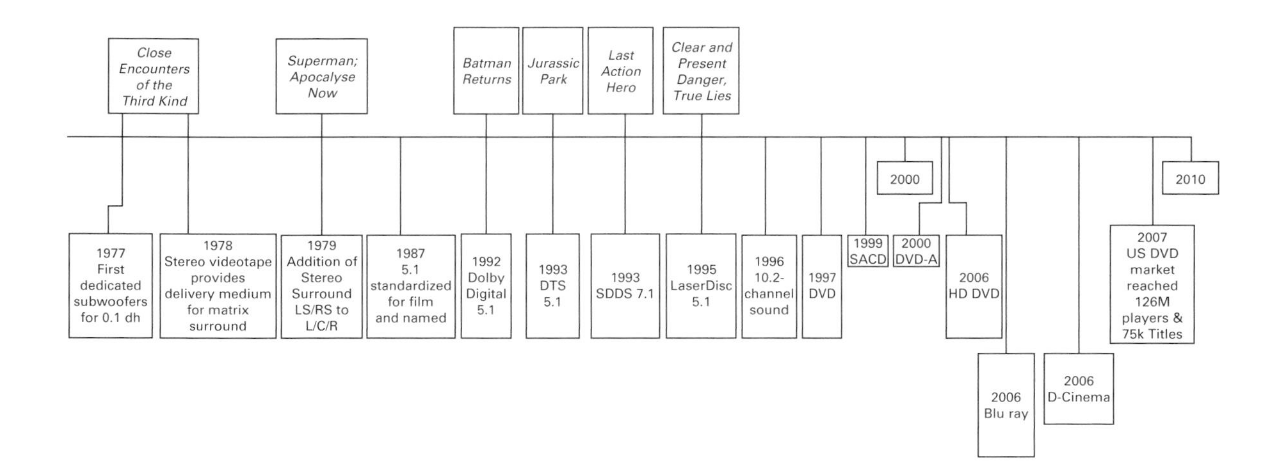
Close Encounters of the Third Kind
Superman; Apocalyse Now
Batman Returns
Jurassic Park
Last Action Hero
Clear and Present Danger, True Lies
2000
2010
1977 First dedicated subwoofers for 0.1 dh
1978 Stereo videotape provides delivery medium for matrix surround
1979 Addition of Stereo Surround LS/RS to L/C/R
1987 5.1 standardized for film and named
1992 Dolby Digital 5.1
1993 DTS 5.1
1993 SDDS 7.1
1995 LaserDisc 5.1
1996 10.2-channel sound
1997 DVD
1999 SACD
2000 DVD-A
2006 HD DVD
2007 US DVD market reached 126M players & 75k Titles
2006 Blu ray
2006 D-Cinema

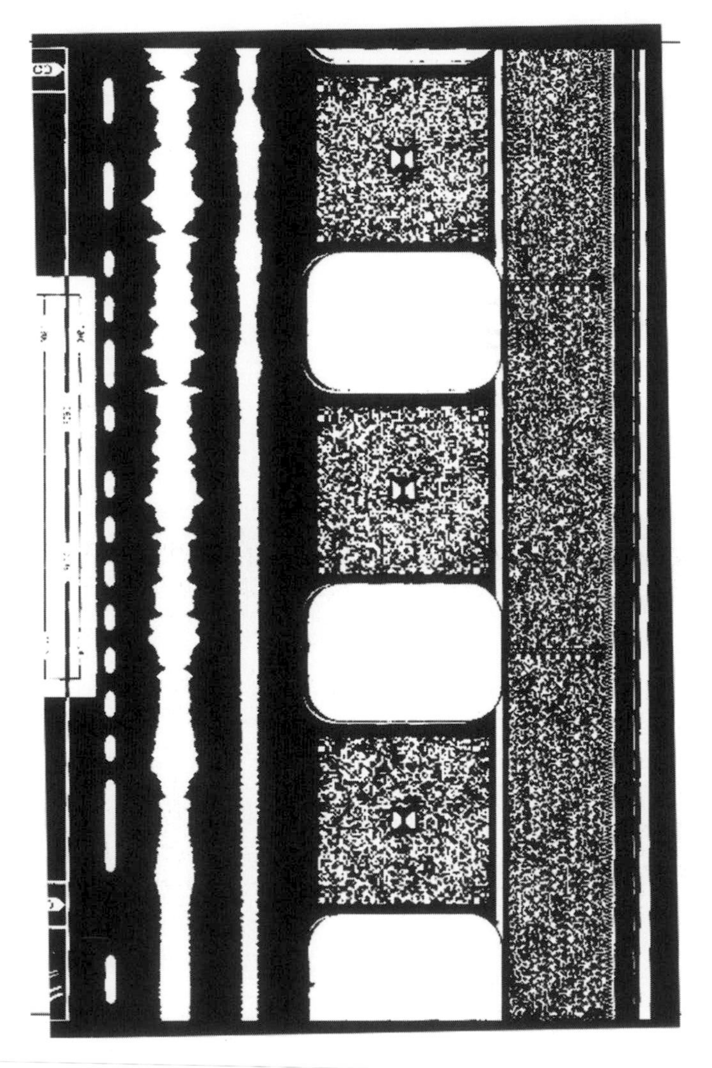

Fig. 1-2 The analog sound track edge of a 35 mm motion picture release print. From left to right is: (1) the edge of the picture area, that occupies approximately four perforations in height; (2) the DTS format time code track; (3) the analog Lt/Rt sound track; (4) the Dolby Digital sound on film track between the perforations; and (5) one-half of the Sony SDDS digital sound on film sound track; the corresponding other half is recorded outside the perforations on the opposite edge of the film.

with normal hearing can hear the difference between mono and stereo, so too can virtually all listeners hear the difference between 2- and 5.1-channel stereo. This process may be seen as one that does have a logical end, when listeners can no longer perceive the difference, but the bounds on that question remain open. Chapter 6 examines the psychoacoustics of multichannel sound, and shows that the pressure upwards in the number of channels will continue into the foreseeable future.

2 Monitoring

Tips from This Chapter

- Monitoring affects the recorded sound as producers and engineers react to the sound that they hear, and modify the program material accordingly.
- Just as a "bright" monitor system will usually cause the mixer to equalize the top end down, so too a monitor system that emphasizes envelopment will cause the mixer to make a recording that tends toward dry, or if the monitor emphasizes imaging, then the recording may be made with excessive reverberation. Thus, monitor systems need standardization for level, frequency response, and amount of direct to reflected sound energy.
- Even "full-range" monitoring requires electronic bass management, since most speakers do not extend to the very lowest audible frequency, and even for those that do, electrical summation of the low bass performs differently than acoustical addition. Since virtually all home 5.1-channel systems employ bass management, studios must use it, at the very least in testing mixes.
- Multichannel affects the desired room acoustics of control rooms only in some areas. In particular, control over first reflections from each of the channels means that half-live, half-dead room acoustics are not useful. Acoustical designers concentrate on balancing diffusion and absorption in the various planes to produce a good result.
- Monitor loudspeakers should meet certain specifications that are given. They vary depending on room size and application, but one principle is that all of the channels should be able to play at the same maximum level and have the same bandwidth.
- Many applications call for direct-radiator surrounds; others call for surround arrays or multidirectional radiators arranged with less direct than reflected sound at listening locations. The pros and cons of each type are given.

- The AES and ITU have a recommended practice for speaker placement in control rooms; cinema practice differs due to the picture. In control rooms center is straight ahead; left and right are at ±30° from center; and surrounds are at ±110° from center, all viewed in plan (from overhead). In cinemas, center is straight ahead; left and right are typically at ±22.5° from center (the screen covers a subtended angle of 50° for Cinemascope in the best seat); and the surrounds are arrays. Permissible variations including tolerances and height are covered in the text.
- So-called near-field monitoring may not be the easy solution to room acoustics problems that it seems to promise.
- If loudspeakers cannot be set up at a constant distance from the principal listening location, then electronic time delay of the loudspeaker feeds is useful to synchronize the loudspeakers. This affects mainly the inter-channel phantom images.
- Low-Frequency Enhancement (LFE), the 0.1 channel, is defined as a monophonic channel having 10 dB greater headroom than any one main channel, which operates below 120 Hz or lower in some instances. Its reason for being is rooted in psychoacoustics.
- Film mixes employ the 0.1 channel that will always be reproduced in the cinema; home mixes derived from film mixes may need compensation for the fact that the LFE channel is only optionally reproduced at home.
- There are two principal monitor frequency response curves in use, the X curve for film, and a nominally flat direct sound curve for control room monitoring. Differences are discussed.
- All monitor systems must be calibrated for level. There are differences between film monitoring for theatrical exhibition on the one hand, and video and music monitoring on the other. Film monitoring calibrates each of the surround channels at 3 dB less than one front channel; video and music monitoring calibrates all channels to equal level. Methods for calibrating include the use of specialized noise test signals and the proper use of a sound level meter.

Introduction

Monitoring is key to multichannel sound. Although the loudspeaker monitors are not, strictly speaking, in the chain between the microphones and the release medium, the effect that they have on mixes is profound. The reason for this is that producers and engineers judge the balance, both the octave-to-octave spectral balance and the "spatial balance" over the monitors, and make decisions on what sounds good based on the representation that they are hearing. The purpose of this chapter is to help you achieve neutral monitoring, although there are variations even within the designation of neutral that should be taken into consideration.

How Monitoring Affects the Mix

It has been proved that competent mixers, given adequate time and tools, equalize program material to account for defects in the monitor system. Thus, if a monitor system is bass heavy, a good mixer will turn the bass down, and compensate the *mix* for a *monitor* fault. Therefore for instance, the monitor system must be neutral, not emphasizing one frequency range over another, and representing the fullest possible audible frequency range if the mix is to translate to the widest possible range of playback situations.

Full-Range Monitoring

Covering the full frequency range is important. In many cases it is impractical to have five "full-range" monitors. Although many people believe that they have "full-range" monitors, in fact most studio monitors cut off at 40 or 50 Hz. Full range is defined as extending downwards to the lowest frequencies audible as continuous sound, say 20 Hz. Thus, many monitors miss at least a whole octave of sound, from 20 to 40 or 50 Hz. The consequence of using monitors that only extend to 40 Hz, say, is that low-frequency rumble may not even be heard by the engineer that is present in the recording. Most home multichannel systems contain a set of electronics called bass management or bass redirection. These systems extract the bass below the cutoff frequency of the 5 channels and send the sum to a subwoofer, along with the content of the Low-Frequency Enhancement (LFE), or 0.1, channel. Therefore, it is common for the professional in the studio that is not equipped with bass management not to hear a low-frequency rumble in a channel, because the monitors cut off at 50 Hz. The listener at home, on the other hand, may hear the rumble, because despite the fact that his home 5-channel satellite loudspeakers cut off at 80 Hz, his bass management system is sending the low-frequency content to a subwoofer that extends the frequency range of all of the channels downward to, say, 30 Hz!

Another fact is that if five full-range monitors are used without bass management to monitor the program production and the program is then played back with bass management at home, electrical summation of the channels may result in phase cancellation that was not noticed under the original monitor conditions. This is because acoustic summation in the studio and electrical summation at home may well yield different results, since electrical summation is sensitive to phase effects in a different way than acoustic summation.

Spatial Balance

The "spatial balance" of sound mixes is affected by the monitor system too, and has several components. One of these is the degree of imaging versus spaciousness associated with a source in a recording. This component is most affected by the choice of microphone and its distance to the source in the recording room, but the perception of space is affected by the monitor and the control room acoustics as well. Here is how: if the monitor is fairly directional, the recording may seem to lack spaciousness, because the listener is hearing mostly the direct sound from the loudspeaker. There is little influence by the listening room acoustics, and the image can seem too sharp—a point source, when it should seem larger and more diffuse. So the recording engineer moves the microphones away from the source, or adds reverberation in the mix. If, on the other hand, the monitor has very wide dispersion and is in a more reverberant space, the mix can seem to lack sharpness in the spatial impression or imaging, and the recording engineer tends to move the microphone in closer, and reduce reverberation in the mix. In both cases, the mix itself has been affected by the directional properties of the monitor loudspeaker (Fig. 2-1).

I had a directly relevant experience in recording Handel's Messiah for the Handel and Haydn Society of Boston some years ago. We set up the orchestra and chorus in the sanctuary of a church, and a temporary control room in a secondary chapel located nearby. We recorded for a day, and then I took the recordings home to listen. The recorded perspective was far too dry when heard at home, and the whole day's work had to be thrown out. What happened was that the reverberation of the rather large chapel was indistinguishable from reverberation in the recording, so I made the recording perspective too dry to compensate for the "control room" reverberation. We moved the temporary control room to a more living room-like space, and went on to make a recording that *High Fidelity* magazine reviewed as being the best at the time in a crowded field. So the monitor system environment, including loudspeakers and room acoustics, affect the recorded sound, because mixers use their ears to choose appropriate balances and the monitor can fool them.

Also, the effects of equalization are profound, and include the apparent distance from the source. We associate the frequency range around 1–3 kHz with the perception of presence. Increasing the level in this region makes the source seem closer, and decreasing it makes the source seem further away. Thus, equalizers in this range are called presence

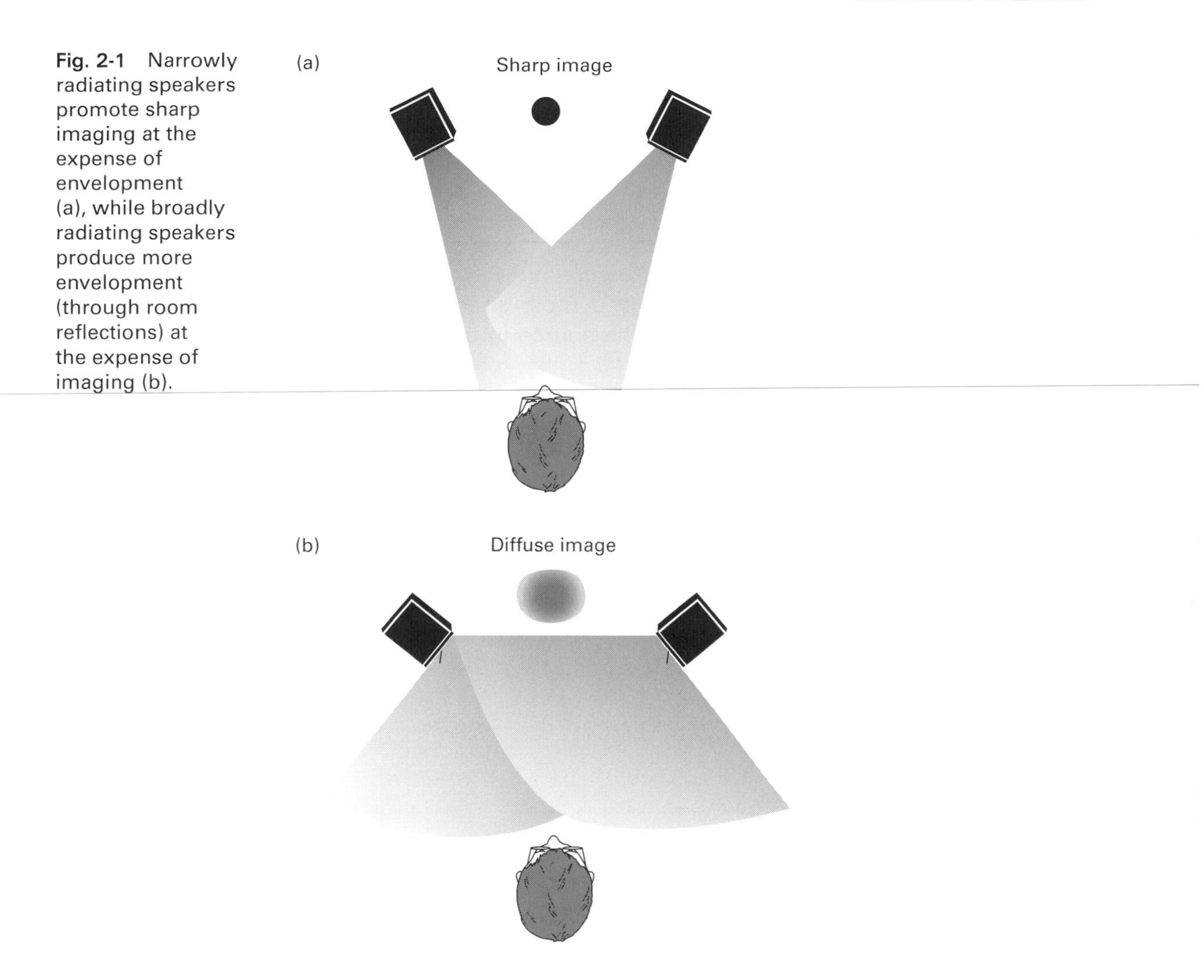

Fig. 2-1 Narrowly radiating speakers promote sharp imaging at the expense of envelopment (a), while broadly radiating speakers produce more envelopment (through room reflections) at the expense of imaging (b).

equalizers, and one console maker has gone so far as to label peaks in this range "presence," and dips "absence." Equalization affects timbre as well, and trying to use the presence range to change apparent distance is not likely to be as effective as moving the microphone, since equalization will have potentially negative effects on the reproduction of naturalness of the source.

In multichannel sound in particular, the directionality of the monitors has a similar effect as in stereo, but the problems are made somewhat different due to the ability to spread out the sound among the channels. For instance, in 5.1-channel sound, reverberation is likely to appear in all 5 channels, or at least in four neglecting the center. After all, good reverberation is diffuse, and it has been shown that spatially diffuse reflections and reverberation contribute to a sense of immersion in a sound field, a very desirable property. The burden imposed on loudspeakers used in 2-channel monitoring to produce both good sound imaging and envelopment at one and the same time is lessened. Thus, it could be argued that we can afford somewhat more directional loudspeakers in multichannel sound than we used for 2-channel sound, because sound from the loudspeakers, in particular the surround loudspeakers, can supply the ingredient in 2-channel stereo that is "missing," namely spaciousness. That is, a loudspeaker that sprays sound around the room tends to produce a greater sense of envelopment through delivering the sound from many reflected angles than does a more directional loudspeaker, and many people prefer such loudspeakers in 2-channel stereo. They are receiving the sensation of envelopment through reflections, while a surround sound system can more directly provide them through the use of the multiple loudspeaker channels. Therefore, a loudspeaker well suited for 2-channel stereo may not be as well suited for multichannel work. Also, for some types of program material it may make sense to use different types of front and surround loudspeakers, as we shall see.

Room Acoustics for Multichannel Sound

Room acoustics is a large topic that has been covered in numerous books and journal articles. For the most part, room acoustics specific to multichannel sound uses the work developed for stereo practice, with a few exceptions. Among the factors that have the same considerations as for stereo are:

- *Sound isolation*: Control rooms are both sources and receivers of sound. The unintentional sound that is received from the outside world interferes with hearing details in the work underway, while that transmitted from the control room to other spaces may be considered noise by those occupying the other spaces. Among considerations in sound isolation are: weight of construction barriers including floor, ceiling, walls, and windows and doors; isolating construction such as layered walls; sealing of all elements against air leaks; removal or control of flanking paths such as over the top of otherwise well-designed wall sections; and prevention or control of noise around penetrations such as wall outlets.
- *Background noise due to HVAC (heating, ventilation, and air conditioning) systems and equipment in the room*: The background noise of 27 home listening rooms averages NC-17. NC means noise criterion curves, a method for rating the interior noise of rooms. NC-17 is a quite low number, below that of many professional spaces. One problem that occurs is that if the control room is noisy, and the end

listener's room is quiet, then problems may be masked in the professional environment that become audible at home. This is partly overcome by the professional playing the monitor more loudly than users at home, but that is not a complete solution. Also, even if the air handling system has been well designed for low noise, and for good sound isolation from adjacent spaces, equipment in the room may often contribute to the noise floor. Computers with loud fans are often found in control rooms, and silent control panels and video monitors wired to operate them by remote control are necessary (Fig. 2-2).

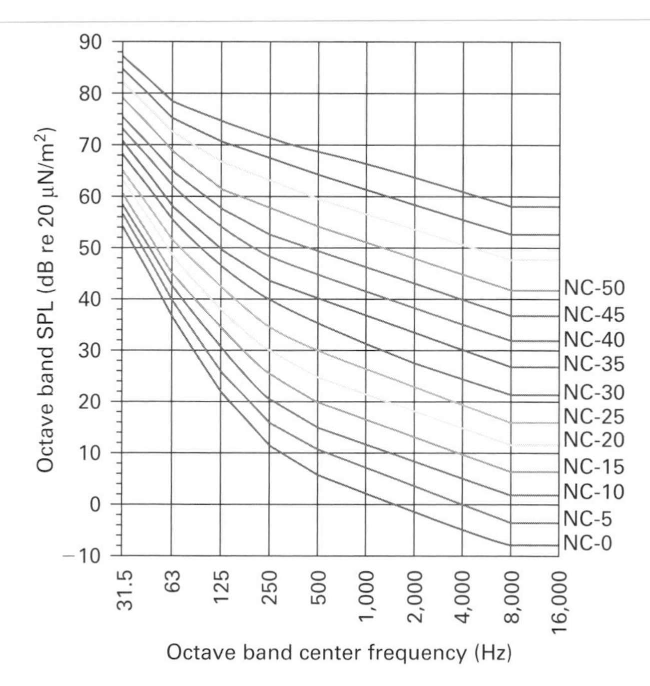

Fig. 2-2 Balanced noise criterion (NC) curves. The background noise of a space is rated in NC units based on the highest incursion into these curves by the measured octave band noise spectrum of the room. The original NC curves have been extended here to the 31.5 Hz and 16 kHz octave bands. An average of 27 home listening rooms measured by Elizabeth Cohen and Lewis Fielder met the curve NC-17, by which is meant that the highest level across the spectrum found just touched the NC-17 curve. Note that a spectrum that follows these curves directly sounds rumbly and hissy, so it is not a design goal.

- *Standing wave control*: This is perhaps the biggest problem with conventional sized control rooms. Frequencies typically from ca. 50 to 400 Hz are strongly affected by interaction with multiple room surfaces, making the distribution of sound energy throughout the room very uneven in this frequency range. Among the factors that can reduce the effects of standing waves are choice of the ratio of dimensions of the room; use of rectangular rooms unless a more geometric shape has been proved through modeling to have acceptable results (a rectangular room is relatively easy to predict by numerical methods, while other shapes are so difficult as to confound computer analysis; in such cases we build a scale model and do acoustic testing of the model); and low-frequency absorption, either through thick absorbing material or through resonant absorbers such as membrane absorbers tuned to a particular frequency range. The number and placement of subwoofers also affects the uniformity of response throughout the space below the bass management crossover frequency, and multiple subwoofers have been shown to be valuable for this reason alone.[1]

The main items that are different for multichannel sound are:

- All directions need to be treated equally: no longer can half-live, half-dead stereo acoustic treatments for rooms be considered usable.
- In the last two decades both the measurement of, and the psychoacoustic significance of, early reflections have become better understood. Early reflections play a different role in concert hall acoustics and live venues than in control rooms. In control rooms, many think that early reflections should be controlled to have a spectrum level less than −15 dB relative to the direct sound for the first 15 ms above 2 kHz (and −20 dB for 20 ms would be better, but difficult to achieve). Note that this is not the "spike" level on a level versus time plot, but rather the spectrum level of a reflection compared to that of the source. Since reflections off consoles in control rooms with conventionally elevated monitor speakers, installed over a control room window can measure −3 dB at 2 ms, it can be seen that such an installation is quite poor at reflection control. Lowering the monitor loudspeakers and having them radiate at grazing incidence over the console meter bridge is better practice, lowering the reflection level to just that sound energy that diffracts over the console barrier, which is much less than the direct reflection from an elevated angle.

Table 2-1 gives a synopsis of the consensus among one group of people for acoustical conditions for multichannel sound. References that can be consulted specific to room acoustics for multichannel sound include ITU-R BS.1116 (www.itu.ch) and EBU Rec. R22 (www.ebu.ch). Reflection control for multichannel control rooms is discussed in "A Controlled-Reflection Listening Room for Multi-Channel Sound" by Robert Walker,

[1] Welti, T. and Devantier, A., "Low-frequency optimization using multiple subwoofers," *JAES*, Vol. 54, No. 5, pp. 347–364, May 2006.

AES Preprint 4645. Note that these standards are conventions. That is, a particular group of engineers has contributed to them to produce what they believe will help to make recordings that are interchangeable among the group of users. As in all such standards these should not be taken as absolute, but rather as an agreement among one group of users.

Table 2-1 International Broadcasters Consensus for Acoustical Conditions for Multichannel Sound

Item	*Specification*
Room size	215–645 ft^2
Room shape	Rectangular or trapezium (in order that the effects of rooms modes be calculable; if a more complex shape is desired, then 1:10 scale model testing is indicated)
Room symmetry	Symmetrical about center loudspeaker axis
Room proportions	$1.1 \times (w/h) < l/h < 4.5 \times (w/h) - 4$ and $l/h < 3$ and $w/h < 3$ where l = length, w = width, and h = height; ratios of l, w, and h that are within ±5% of integer values should be avoided
RT60, 200 Hz to 4 kHz average	$T_m = 0.3 \times (V/V_o)^{1/3}$ with a tolerance of +25−0% where V = volume of room and V_o = the reference volume of 100 m^3 (3,528 ft^3)
Diffusion	Should be high and symmetrically applied (No. 1/2 live, 1/2 dead); no hard reflecting surfaces returning sound to critical listening areas >10 dB above level of reverberation (ports, doors special concern)
Early reflection control	−15 dB spectrum level for first 15 ms
Background noise	<NC-17 US based on survey of many living rooms, NR15 European equivalent

A similar agreement is ISO 2969 for cinemas, which standardizes the "X" curve. Because cinemas use similar directivity speakers in similar acoustic environments, interchangeability of program material is enhanced when such a single curve is employed, but although thoroughly standardized many users may not realize that it is in fact a series of responses with a dependency on room size. The wider range of acoustical conditions for professional versus home listening mean that interchangeability is lessened, and exact matches, which are sought on the professional level, are unlikely.

Note that reverberation time is a concept that depends in general on their being a diffuse sound field. While true of large reverberant or so-called Sabine spaces, named for Wallace Clement Sabine the father of modern acoustics, small rooms do not in general produce a classic reverberant sound field, but rather the factor called reverberation time is actually dominated by the decay time of the various standing waves, especially throughout the important mid-bass region. Thus equations used to predict reverberation time are often faulty in predicting the decay time of small rooms, since the mechanisms for reverberant decay and modal decay are different (Fig. 2-3).

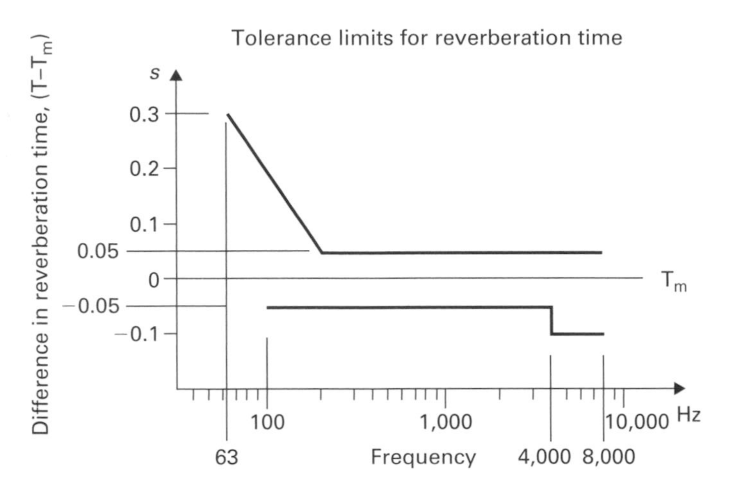

Fig. 2-3 The tolerance for reverberation time in terms of permissible difference in RT60 from the time given in Table 2-1 for a specific room volume, as a function of frequency.

A different view of some of these specifications is held by a prominent researcher, Dr. Floyd Toole. In his paper "Loudspeakers and Rooms for Sound Reproduction—A Scientific Review," *AESJ*, Vol. 54, No. 6, pp. 451–476, he argues for the benefits of early reflections, among many other things. Since this work is relatively new, it has not been absorbed into standards yet. However, I can state that in my experience we do adapt to a stronger set of early reflections in one room versus another, because among other places I work in a space designed in the early 1980s that exhibits strong early reflections. Upon entering the space, especially from listening in a more reflection controlled space, I find the timbral signature of comb filtering obvious, yet within a few minutes I adapt to this space and can perform program equalization about as well as in the deader space. A cinema that employs diffuse early reflections at a higher level than normal from the side walls has been built in Europe, and enjoys a good reputation. While staying within the maximum reverberation time set by the THX standards that I wrote, this cinema shows that there may be an advantage to diffuse side wall reflections.

Another observation: putting a person going "tsss, tsss, tsss, tsss" on the conductor's podium at Symphony Hall Boston and rotating around one finds listening in mid-audience that the side wall reflections are very prominent, almost what one would call an echo. This has been known for some time to be a distinguishing feature of shoe-box shaped halls making them more desirable than fan-shaped halls on music, although heard this way it seems like a defect. How this concert hall finding makes its way into small room acoustics is unclear however since the time frame of reflections is so much shorter in small than in large rooms.

Choice of Monitor Loudspeakers

The choice of monitor loudspeakers depends on the application. In larger dubbing stages for film and television sound, today's requirements for frequency range, response, and directivity are usually met by a combination of direct radiating low-frequency drivers and horn-loaded high-frequency compression drivers. In more conventionally sized control rooms, direct radiating loudspeakers, some of which are supplied with "waveguides" that act much like horns at higher frequencies, are common.

The traditional choice of a control room monitor loudspeaker was one that would play loudly enough without breaking. In recent years, the quality of monitor loudspeakers has greatly improved, with frequency range and response given much more attention than in the past. So what constitutes a good monitor loudspeaker today?

- Flat and smooth frequency response over an output range of angles called a listening window. Usually this listening window will be an average of the response at points on axis (0°), and at ±15° and ±30° horizontally and ±15° vertically, recognizing the fact that listeners are arrayed within a small range of angles from the loudspeaker, and that this range is wider horizontally than vertically. By averaging the response at some seven positions, the effects of diffraction off the edges of the box and reflections off small features such as mounting screws, which are not very audible, are given due weight through spatial averaging.
- A controlled angle of the main output versus frequency. This is a factor that is less well known than the first, and rarely published by manufacturers, but it has been shown to be an important audible factor in both loudspeaker design and interaction with room environments. A measure of the output radiation pattern versus frequency is called the directivity index, DI, and it is rated in dB, where 0 dB is an omnidirectional radiator, 3 dB a hemispheric radiator, 6 dB a quarter-sphere radiator, and 9 dB an eighth-sphere radiator.

  What DI to use for monitor loudspeakers has been the subject of an on-going debate. Some experimental work on the subject was done for 2-channel stereo in the 1970s, and it showed "changes in mid-frequency directivity of about 3 dB were very noticeable due to the change in definition, spatial impression, and presence ... The results ... were that a stereo loudspeaker should have a mid-frequency directivity of about 8 dB with a very small frequency dependency." However, this is an old paper and the results have not been followed up on. It is not reflected in contemporary designs because achieving directivity this high in the mid-range is difficult. Typical good-quality monitor speakers today have more like 4–5 dB DI through the mid-range.

Abrupt changes in DI across frequency cause coloration, even if the listening window frequency response is flat. For instance, if a two-way speaker is designed for flat, on-axis response, and crosses over at a frequency where the woofer's radiation pattern is narrow to a much wider radiating tweeter, the result will be "honky" sounding. This is rather like the sound resulting from cupping your hands to form a horn in front of your mouth, and speaking. On the other hand, it is commonplace for loudspeakers to be essentially omnidirectional at low frequencies, increasing smoothly in the mid-range, and then increasing again at the highest frequencies. It appears to be best if the DI can be kept reasonably smooth although not constant across the widest frequency range possible—this means that key first reflections in the environment are more likely to show a smooth response.

Some loudspeaker designs recognize the fact that in many typical listening situations the first reflections from the ceiling and floor are the most noticeable, and these designs will be made more directional in the vertical plane than the horizontal, through the use of horns, arrays of cone or dome drivers, or aperture drivers (such as a ribbon).

- Adequate headroom. In today's digital world, the upper limit on sound pressure level of the monitor is set by a combination of the peak recordable level, and the setting of the monitor volume control. The loudspeaker should not distort or limit within the bounds established by the medium, reference volume control setting, and headroom. Thus, if a system is calibrated to 78 dB SPL for −20 dBFS on the medium, the loudspeaker should be able to produce 98 dB SPL at the listening position, and more to include the effects of any required boost room equalization, without audible problems.
- Other factors can be important in individual models, such as signal-to-noise ratio of internal amplifiers, distortion including especially port noise complaints, and the like.

- There are three alternatives for surround loudspeakers: conventional direct radiators matching the fronts, surround arrays, and multidirectional loudspeakers (Table 2-2). Surround arrays and multidirectional designs are covered later in this chapter.

Table 2-2 Loudspeaker Specifications for Multichannel Sound

Direct radiating loudspeaker specifications	***Applies to front, and one type of surround speaker***
Listening window frequency response, average of axial and ±15° and ±30° horizontally, and ±15° vertically	From subwoofer crossover frequency to 20 kHz, ±2 dB, with no wide-range spectral imbalance
DI	Desirable goal is smoothly increasing directivity over frequency. Practical loudspeakers currently exhibit 0 dB at low frequencies increasing to 6–8 dB from 500 Hz to 10 kHz with a tolerance of ±2 dB of the average value in this range, then rise at higher frequencies.
THD, 90 dB SPL <250 Hz	Not over −30 dB
THD, 90 dB SPL ≥400 Hz	Not over −40 dB
Group delay distortion	<0.5 ms at 200 Hz to 8 kHz, <3 ms at 100 Hz and 20 kHz
Decay time to 37% output level	t < 5/f, where f is frequency
Clipping level	Minimum 103 dB SPL at listening position, but depends on application, and more should be added for equalization. Can be tested with "boinker" test signal available on the test CDs.*
Widely dispersing loudspeaker specifications such as multidirectional	*For surround use only; an alternate type to direct radiator*
Power response	±3 dB from subwoofer crossover to 20 kHz
Directivity	Broad null in the listening direction is typical
Other characteristics except frequency response and directivity	Equal to direct radiator
Subwoofer	
Frequency response measured including low-frequency room gain effects	±2 dB, 20 Hz to crossover frequency
Power handling	Should handle maximum level of all 5.1 channels simultaneously, 18 dB greater than the maximum level of 1 channel
Distortion	All forms of distortion (harmonic, inharmonic, intermodulation, and noise-like distortions) should be below human masking thresholds; this will ensure inaudible distortion and make localizing the subwoofer unlikely
Group delay	<5 ms difference, 25 Hz to subwoofer crossover

*http://www.hollywoodedge.com/product1.aspx?SID=7sProduct_ID=951058Category_ID=12063

One Standardized Setup

One standardized setup for 5.1-channel sound systems is that documented by the AES in its document TD-1001[2] and by the International Telecommunications Union (ITU), in their recommendation 775. In this setup, the speakers are all in a horizontal plane that matches your ear height, or are permitted to be somewhat elevated if that is necessary to provide a clear path from the loudspeaker to the listener. Center is, of course, straight ahead, at 0° from the principal listening location (Table 2-3).

Table 2-3 Loudspeaker Locations for Multichannel Sound Accompanying a Picture

Front loudspeaker location	Centerline of the picture for the center loudspeaker. At edges of screen just inside or outside picture depending on video display. 4° maximum error between picture and sound image in horizontal plane. Height relationship to picture depends on video display and number of rows of seating, etc.
Surround loudspeaker location	±110° from center in plan view with a tolerance of ±10° and at seated ear height minimum or elevated up to 30°
Subwoofer(s)	Located for best response

Left and Right

Left and right front speakers are located at ±30° from center, when viewed from above, in plan. This makes them 60° apart in subtended angle. Sixty degrees between left and right, forming an equilateral triangle with the listener, has a long history in 2-channel stereo, and is used in 5.1-channel stereo for many of the same reasons that it was in 2-channel work: wider is better for stereo perception, while too wide makes for problems with sound images lying between the channels. While the center channel solves several problems of 2-channel stereo including filling in the "hole" between left and right (see Chapter 6), it has nevertheless been found by experimenters that maintaining the 60° angle found commonly in 2-channel work is best for 5.1-channel sound (Fig. 2-4).

There is a further consideration in left/right angles when sound is to accompany a picture described on page 42.

[2]http://www.aes.org/technical/documents/AESTD1001.pdf

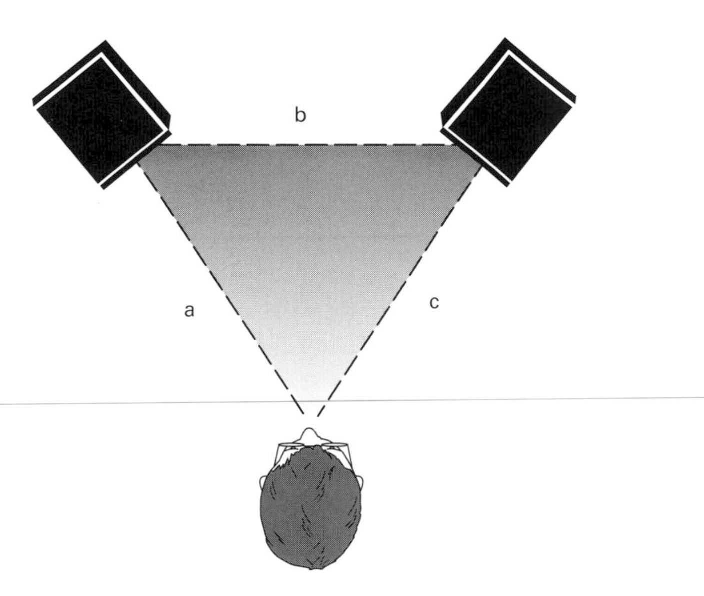

Fig. 2-4 In the standard stereo setup, a = b = c and an equilateral triangle is formed. The listener "sees" a subtended angle of 60° from left to right.

Surround

The surround loudspeakers in the plan are located at ±110° ±10° from front center, that is, between 100° and 120° from front center. This angle was determined from experiments into reproduction of sound images all around versus producing best envelopment of the listener. More widely spaced (further from the front) surrounds produced better rearward sound images at some expense in envelopment, while further forward surrounds produce better envelopment at the expense of rearward imaging. Also, 110° reportedly better represents the likely home listening situation where the principal listening position is close to a rear wall of the space, rather than in the middle of the space if the loudspeakers were more widely placed.

Surround loudspeakers height is often elevated to avoid control room equipment, doors, and so forth, and this is permissible for many kinds of surround presentations. In cinema usage and in many home systems, surrounds are elevated with respect to the audience. However, in certain types of presentations, such as music with a "middle of the band" perspective, the elevated surround may cause a curious tilt in the resulting sound field, so may not be as well liked. Applause reproduced with some kinds of live programming is also anomalous as it does not correspond to what is actually found in halls. Thus surround elevation is a question the answer to which depends largely on the program material being made in a particular studio destined for a particular reproduction situation.

Subwoofer

The subwoofer in a bass managed system carries the low-frequency content of all 5 channels, plus the 0.1 LFE channel content; this is a consideration in the placement. Among the others are:

- Placement in a corner produces the most output at low frequencies, because the floor and two walls serve as reflectors, increasing the output through "loading"; the subwoofer may be designed for this position and thus reduce the cone motion necessary to get flat response and this improves low-frequency headroom.
- Making an acoustical splice between the subwoofer and each of the channels is a factor that can be manipulated by moving the subwoofer around while measuring the response.
- The placement of the subwoofer and the listener determine how the sound will be affected by standing waves in the room; moving the subwoofer around for this effect may help smooth the response as well.
- Multiple subwoofers placed differently with respect to room boundaries can help smooth the modal response through multiple driving points and thus differing transfer functions between each of them and the listener. Where a room may produce ±12 dB variations in response measured with high-frequency resolution equipment, that variation can be reduced with multiple subs.

Setting Up the Loudspeaker Locations with Two Pieces of String

Many times a trigonometry book is not at hand, and you have to set up a surround monitor system. Here is how:

- Determine the distance from each of the left and right loudspeakers to the principal listening position to be used. Home listening is typically done at 10 ft (3 m), but professional listening is over a wide range due to differing requirements from a large scoring stage control room to a minimum sized booth in a location truck.
- Cut a piece of string to the length of the listening distance, or use a marked XLR extension, and make the distance between the left and right loudspeakers equal to the distance between each of them and the prime listening location. This sets up an equilateral triangle, with a 60° subtended angle between left and right loudspeakers.
- Place the center speaker on the centerline between left and right speakers. Use the full string length again to set the distance from the

listener to the center speaker, putting the front three loudspeakers on an arc, unless an electronic time delay compensation is available (discussed below).

- Use two strings equal in length to the listening distance. Place one from the listener to the left loudspeaker, and the other perpendicular to the first from the listener's location and to the outside of the front loudspeakers. Temporarily place a surround loudspeaker at this location, which is 90° from the left loudspeaker; 90° plus the 30° that the left is from center makes 120°. This angle is within the tolerance of the standard, but to get it right on, swing the loudspeaker along an arc towards the front by one-third of the distance between the left and center loudspeaker. This places it 110° from center front, assuming all the loudspeakers are at a constant distance from the listener.
- Repeat in a mirror image for the right surround channel (Fig. 2-5).

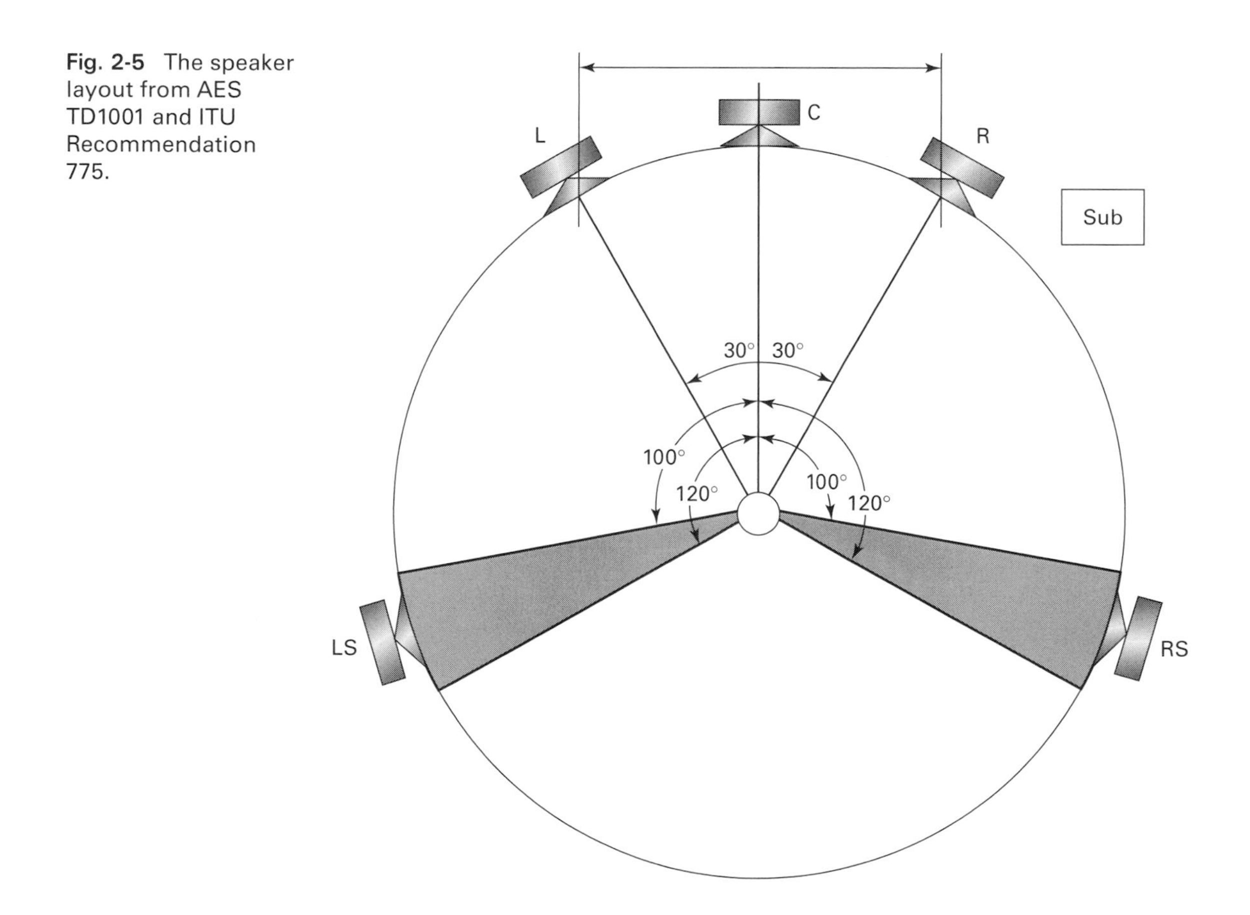

Fig. 2-5 The speaker layout from AES TD1001 and ITU Recommendation 775.

Setup Compromises

Frequently, equipment, windows, or doors are just where the loudspeakers need to be placed. Following are compromises that can be made if necessary:

- Generally, in front hearing is about three times less sensitive to errors in elevation than to horizontal errors, while at the sides and back these listening errors become much larger and therefore freedom in placement is greater. Therefore, it is permissible to elevate loudspeakers above obstructions if necessary. It is best to elevate the loudspeakers only enough to clear obstructions, since if they are too high, strong reflections of sound will occur off control surfaces at levels that have been found to be audible.
- The tolerance on surround loudspeaker placement angles is wide. Probably both surround loudspeakers should use close to the same angle from center, although the range is ±10° from the ±110° angle. Also, surround loudspeaker placement is often a compromise, because if the producer sits behind the engineer, then the surround angles for the producer are significantly less than for the engineer, especially in smaller rooms. With this in mind, it may be useful to use a somewhat wider angle from the engineer's seat than 110° to get the producer to lie within the surrounds, instead of behind them.
- With a sufficiently low crossover frequency and steep filter, along with low distortion from the subwoofer, placement of the subwoofer(s) becomes non-critical for localization. Thus, it or they may be placed where the smoothest response is achieved. This often winds up in a front corner, or if two are used, one in a front corner and one halfway down a side wall, or centered along the two side walls, to distribute the driving points of the room and "fill in" the standing wave patterns.

Center

In many listening situations it may be impossible to place the center loudspeaker at 0° (straight ahead) and 0° elevation. There is probably other equipment that needs to be placed there in many cases. The choices, when confronted with practical situations involving displays or controls, are:

- above the display/control surfaces,
- below the display/control surfaces,
- behind the display.

Placement above the display/control surface is common in many professional facilities, but such placement can carry a penalty in many instances the sound splash off the control surface is strongly above audibility. Recommended placement for monitor loudspeakers is to set them up so that the principal listener can see the whole speaker, just over the highest obstruction. This makes the sound diffract over the obstruction, with the obstruction, say a video monitor, providing an acoustic shadow so far as the sound control surface is concerned. The top of the obstruction can be covered in thin absorbing material to absorb the high frequencies that would otherwise reflect off the obstruction.

Another reason not to elevate the center speaker too much is that most professional monitor loudspeakers are built with their drivers along a vertical line, when used with the long dimension of the box vertically. This means that their most uniform coverage will be horizontal, and vertically they will be worse, due to the crossover effects between the drivers. Thus, if a speaker is highly elevated and tipped down to aim at the principal mixer's location, the producer's seat behind the mixer is not well covered—there are likely to be mid-frequency dips in the direct-field frequency response. (The producer's seat also may be up against the back wall in tight spaces, and this leads to more bass there than at the mixer's seat.)

A third reason not to elevate the studio monitor loudspeaker too much is that we listen with a different frequency response versus vertical angle. Called head-related transfer functions (HRTFs), this effect is easily observed. Playing pink noise over a monitor you are facing, tip your head up and down; you will hear a distinct response change. Since most listeners will not have highly elevated speakers, we should use similar angles as the end user in order to get a similar response.

A position below the display/control surface is not usable in most professional applications, for obvious reasons, but it may be useful in screening rooms with direct view or rear-projection monitors. The reason that this position may work is that people tend to locate themselves so they can see the screen. This makes listeners that are further away elevated compared to those closer. The problem with vertical coverage of the loudspeaker then is lessened, because the listeners tend to be in line with each other, viewed from the loudspeaker, occupying a smaller vertical angle, with consequently better frequency response.

When it is possible, the best solution may well be front projection with loudspeakers behind the screen using special perforated screens. Normally perforated motion picture theater screens have too much high-frequency sound attenuation to be useful in video applications, but in the last few years several manufacturers have brought out screens with much smaller perforations that pass high-frequency sound with near transparency, and that are less visible than the standard perforations. However, perforation patterns may interact with the pixel pattern in modern projectors and produce moiré patterns. For such cases, the woven and acoustically transparent screens become a necessity, such as those of Screen Research.

By the way, it is not recommended that the center loudspeaker be placed "off center" to accommodate a video monitor. Instead, change the loudspeaker elevation, raising it above the monitor, and moving it back as shown in the Fig. 2-6, creating an "acoustic shadow" so the effect of the direct reflection off the console is reduced. This is permissible since listening is more sensitive to errors in the horizontal plane than in the vertical one.

A "top and bottom" approach to the center channel to try to center a vertical phantom image on the screen may have some initial appeal, but it has two major problems. The first is that at your eardrum these two have different responses, because of the different HRTFs at the two angles. The second is that then you become supremely sensitive to seated ear height, as the change of an inch off the optimum could bring about an audible comb filter notch at high frequencies.

Left and Right

One problem that occurs when using sound accompanying a picture with the ±30° angle for left and right speakers is that they are unlikely to be placed inside the picture image area using such a wide angle. Film sound relies on speakers just inside the left and right extremities of the screen to make front stereo images that fit within the boundaries of screen. The reason for this is so that left sound images match left picture images and so forth across the front sound field. But when film is translated to video, and the loudspeakers are outside instead of inside the boundaries of the screen, some problems can arise. Professional listeners notice displacement between picture and sound images in the horizontal plane of about 4°, and 50% of the public becomes annoyed when the displacement reaches 15°. So, in instances where the picture is much smaller than 60°, there may need to be a compromise between what is best for sound (±30°), and what is best for sound plus picture (not much wider than 4° outside the limits of the picture). For those mixing program content with no picture, this is no consideration, and there are programs even with a picture where picture–sound placement is not so important as it is with many movies. Thus for a 32° total horizontal subtended angle

Fig. 2-6 When an associated video picture or a computer monitor are needed, the alternatives for center are above, below, or behind the associated picture, each of which has pros and cons.

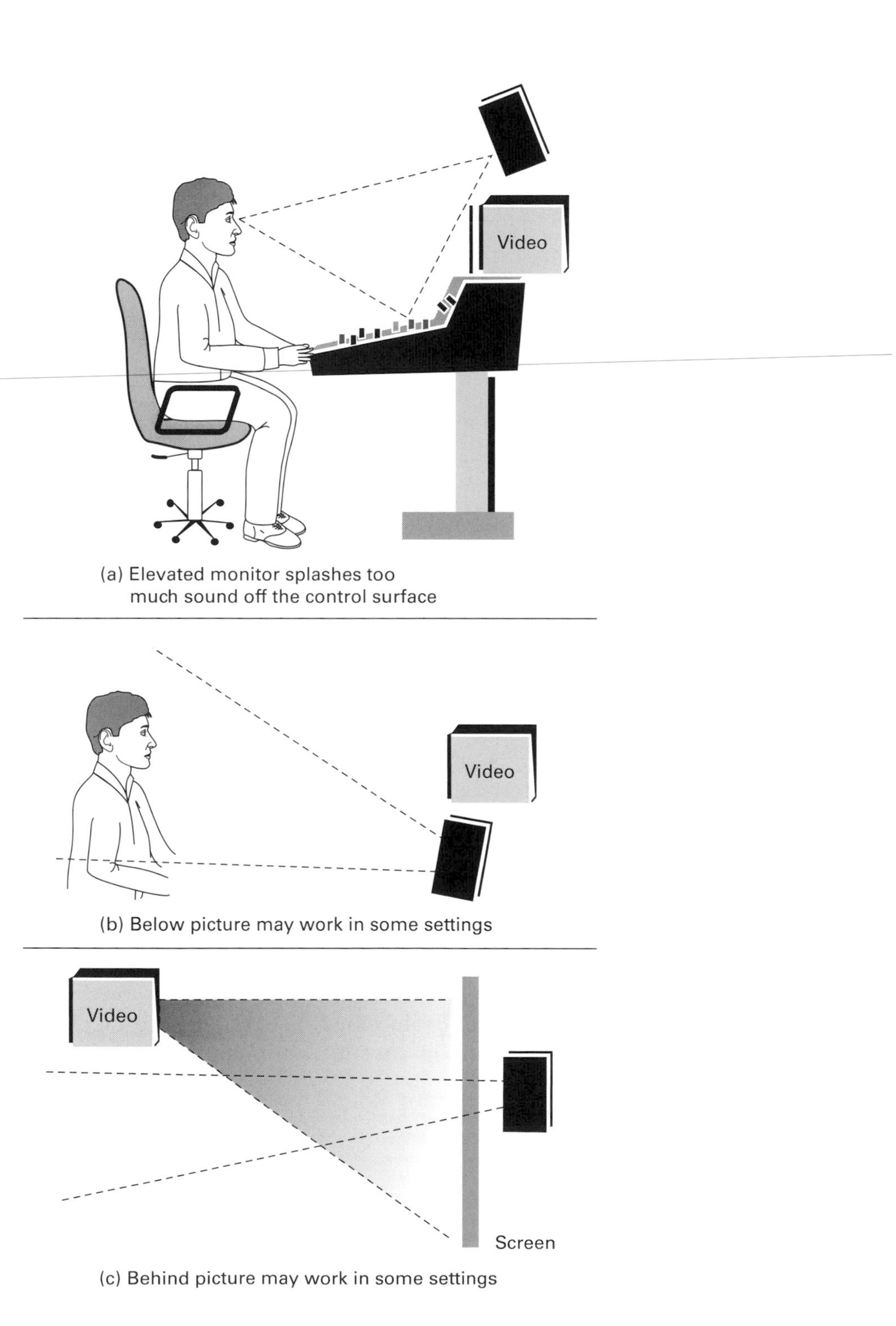

(a) Elevated monitor splashes too much sound off the control surface

(b) Below picture may work in some settings

(c) Behind picture may work in some settings

high-definition screen,[3] speakers at ±30° are seen as too wide, and they would normally be brought in to be just outside the picture (or just within it if front projection and perforated screens are in use). This does limit auditory source width of direct channels and so is a compromise for sound, but is necessary for conditions where picture and sound must match localization.

Surround

Even if the surround loudspeakers are the same model as the fronts, there will still be a perceived frequency response difference. This is due to the HRTFs, the fact that the frequency response in your ear canal determines the spectrum that you hear, not the frequency response measured with a microphone. Your head has a different response for sound originating in front compared to the rear quadrant, and even when the sound fields are perfectly matched at the position of the head, they will sound different. A figure in Chapter 6 shows the frequency response difference in terms of what equalization has to be applied to the surround loudspeaker to get it to match spectrum with the center front. (This response considers only the direct sound field, and not the effects of reflections and reverberation, so practical situations may differ.)

The surrounds may be elevated compared to the fronts without causing much trouble. As they get more overhead, however, they may become less distinguishable from one another and thus more like a single monophonic channel, so too high an angle is not desirable. Experiments into surround height reveal little difference from 0° elevation to 45° elevation for most program material. Some mixers complain, however, about elevated monitors if the program contains audience sound like applause: they don't like the effect that the listeners seem located below the audience in these cases.

Subwoofer

In one case, for FCC mandated listening tests to low-bit-rate codecs for Digital Television, I first set up two subwoofers in between the left and center, and center and right loudspeakers. Since we had no means to adjust the time delay to any of the channels (discussed below), I felt that this would produce the best splice between the main channels and the subwoofer. Unfortunately, these positions of driving the room with the subwoofers, which were set up symmetrically in the room, produced lumpy frequency response. I found that by moving one of

[3] This angle is set by noting HDTV has 1920 pixels horizontally, and that 20/20 vision corresponds to an acutance of 1/60° of arc. Dividing 1920 by 60 yields 32°.

the subs one-half way between the front and surround loudspeakers, the response was much smoother. This process is called "placement equalization," and although it requires a spectrum analyzer to do, it is effective in finding placements that work well.

The use of a common bass subwoofer for the 5 channels is based on psychoacoustics. In general, low-frequency sound is difficult to localize, because the long wavelengths of the sound produce little difference at the two ears. Long-wavelength sound flows freely around the head through diffraction, and little level difference between the ears is created. There is a time difference, which is perceptible, but decreasingly so at lower frequencies. This is why most systems employ five limited bandwidth speakers, and one subwoofer doing six jobs, extending the 5 channels to the lowest audible frequencies, and doing the work of the 0.1 channel. The choice of crossover frequency based on finding the most sensitive listener among a group of professionals, then finding the most sensitive program material, and then setting the frequency at two standard deviations below the mean found by experiment, resulted in the choice of 80 Hz for high-quality consumer systems.

Another factor to consider when it comes to localizing subwoofers is the steepness of the filters employed, especially the low-pass filter limiting the amount of mid-range sound that reaches the subwoofer. If this filter is not sufficiently steep, such as 24 dB/octave, even if it is set to a low-frequency, higher-frequency components of the sound will come through the filter, albeit attenuated, and still permit the subwoofer to be localized. The subwoofer may also be localized in two accidental ways: through distortion components and through noise of air moving through ports. Both of these have higher-frequency components, outside the band of the subwoofer, and may localize the loudspeaker. Careful design for distortion, and locating the speaker so that port noise is directed away from direct listening path, are helpful.

No one has suggested that each of the 5 channels must be extended down to 20 Hz individually. The problem with doing this in any real room is that the low frequency response would vary dramatically from channel to channel, due to the different driving points of the room. Remember that even "full-range" professional monitors have a cutoff of 40–50 Hz, so using bass management below this frequency is still valuable. With very large PMC monitors at CES some years ago, I found the best splice to the Whise 616 subwoofer (a 5-ft cube with 4 15-in. drivers which had a low-frequency rolloff of −1 dB at 16 Hz), to be at 25 Hz! And switching on and off the subwoofer for the 16–25 Hz content was audible on some of the concert hall recorded program material! But the marketplace remains practical, and the best cinema subs usually are down 1 dB at around 24–26 Hz.

Setup Variations

Use of Surround Arrays

In motion picture use, surround arrays are commonplace, having been developed over the history of surround sound from the *Fantasia* onwards. The AES/ITU recommendation recognizes possible advantages in the use of more than two surround loudspeakers, in producing a wider listening area and greater envelopment than available from a pair of direct radiators. The recommendation is made that if they are used, there should be an even number disposed in left and right halves in an array that occupies the region between ±60° and ±150° divided evenly and symmetrically placed. Thus if there are four surround loudspeakers, the pairs would be placed at ±60° and ±150° from center; with six speakers, the pairs would be placed at ±60°, ±105°, ±150°, etc.

However, use of this rather sparse array, with just four or six speakers depends on something vital: uncorrelated signals available for each of them. The original experiment on which the recommendation was made used four surround channels, all the way from different diffuse-field dominant microphones to the loudspeakers. I have found that an array of four surround speakers, with two driven in parallel on each side as one would do with 5.1-channel sources, is particularly poor at coloration because the comb filter fingerprint is so very audible. In small low-reverberation-time rooms for film sound mixing I prefer to use a large number of smaller loudspeakers. In one installation we have 16 surround loudspeakers, six on a side and four across the back, and this works well. These are quite small two-way designs you can hold in your hand, but the array of them sounds better, and the sound is more uniform from front to back, than any point source large surround speaker would be.

Surround arrays have some advantages and disadvantages but are commonplace in large theater spaces. Since they are used there, they also appear in dubbing stages for film, and also for television work that is done on the scale of film, such as high-end television post-production for entertainment programming. The advantages and disadvantages are:

- In large rooms, an array of loudspeakers can be designed to cover an audience area more uniformly, both in sound pressure level and in frequency response, than a pair of discrete loudspeakers in the rear corners of the auditorium. This principle may also apply to smaller control rooms, some of which use arrays.
- In addition it is possible to taper the output of the array so that the ratio of the front channel sound to surround sound stays more

constant from front to back of the listening area (i.e., putting more sound level into the front of the listening space than the back, in the same proportion that the front loudspeakers fall off from front to back helps uniformity of surround impression, the fall off being on the order of 4dB in well-designed theatrical installations). This is an important consideration in making the surround experience uniform throughout a listening space since it is the *ratio* of front to surround sound that is more important than the absolute level of each.

- In the context of sound accompanying a picture, it is harder to localize an array than discrete loudspeakers due to the large number of competing sources, thus reducing the exit sign effect. This effect is due to the fact that when our attention is drawn off the screen by a surround effect, what we are left looking at is not a continuation of the picture, but rather, the exit sign.
- A drawback is that the surround sound is colored by the strong comb filters that occur due to the multiple times of arrival of each of the loudspeakers at listener's locations. This results in a timbral signature rather like speaking in a barrel that affects the surround sound portion of the program material, but not the screen sound part. Noise-like signals take on a different timbre as they are panned from front to surround array. It turns out to be impossible to find an equalization that makes timbre constant as a sound is panned from the screen to the surrounds.
- Another drawback is that pans from the front to the surrounds seem to move from the front to the sides, and not beyond the sides to behind. Discrete rear loudspeakers or 7.1–10.2 channel systems can do this better.

The standards referred to above were written before either matixed 6.1 or discrete 6.1-channel systems separating the back from side surrounds was conceived. However, it is worth noting that in Cinerama, the surround array switch between the two available tracks between left-right surround and front-back surround. This was done manually by the theater projectionist, presumably between the segments in travelogues like *This Is Cinerama*. Surround arrays in cinemas today are divided into four groups: left, right, left back, and right back. The left and left back are driven together for 5.1, as are the right and right back, and the back is separated out for those mixes that have a back channel so that the array becomes left, back, right. The advantage this has is particularly in pans from screen to off screen. With conventional left-right only surround and an array the mixer will find that the sound only images from the screen to the sides of the listener; it can't go behind him.

Surround Loudspeaker Directivity

Depending on the program material and desires of the producer, an alternate to either a pair of discrete direct radiators, or to surround arrays, is to use special radiation pattern surround loudspeakers. A pair of dipole loudspeakers arranged in the AES/ITU configuration, but with the null of the radiation pattern pointed at the listening area, may prove useful. The idea is to enhance envelopment of the surround channel content, as opposed to the experience of imaging in the rear. These work well in the rooms for which they were designed: consumer homes without particular acoustic treatment. In lower-reverberation-time environments for their volume, such as small professional rooms, these are clearly not what is intended. Pros and cons of conventional direct radiators compared to multiradiator surrounds is as follows:

Pros for direct radiators for surround:

- rear quadrant imaging is better (side quadrant imaging is poor with both systems for reasons explained in Chapter 6);
- localization at the surround loudspeaker is easily possible if required;
- somewhat less dependence on room acoustics of the control room.

Cons for direct radiators for surround:

- too often the location of the loudspeakers is easily perceived as the source of the "surround" sound;
- pans from front to surround first snap part of the spectrum to the surround speaker, then as the pan progresses, produces strongly the sound of two separate events, then snaps to the surround; this occurs due to the different HRTFs for the two angles of the loudspeakers to the head, the different frequency response that appears in the ear canal of listeners even if the loudspeakers are matched.

Pros for multidirectional radiators for surround:

- delivers the envelopment portion of the program content (usually reverberation, spatial ambience) in a way that is more "natural" for such sound, that is, from a multiplicity of angles through reflection, not just two primary locations;
- produces more uniform balance between front channel sound and surround sound throughout a listening area; in a conventional system moving off center changes the left-right surround balance much more quickly than with the dipole approach;
- makes more natural sounding pans from front to surround sound, which seem to "snap" from one to the other less than with the direct-radiator approach.

Cons for multidirectional radiators for surround:

- not as good at rear quadrant imaging from behind you as direct radiators;
- localization at the surround loudspeaker location is difficult (this can also be viewed as a pro, depending on point of view—should you really be able to localize a surround loudspeaker?);
- greater dependence on room acoustics of the control room, which is relied upon to be the source of useful reflections and reverberation.

There has been a great deal of hand-wringing and downright misinformation in the marketplace over the choice between direct radiator and multidirectional radiators for surround. In the end, it has to be said that both types produce both direct sound and reflected sound, so the differences have probably been exaggerated (Fig. 2-7). (Multidirectional radiators produce "direct sound" not so much by a lack of a good null in the direction of the listener as from discrete reflections.)

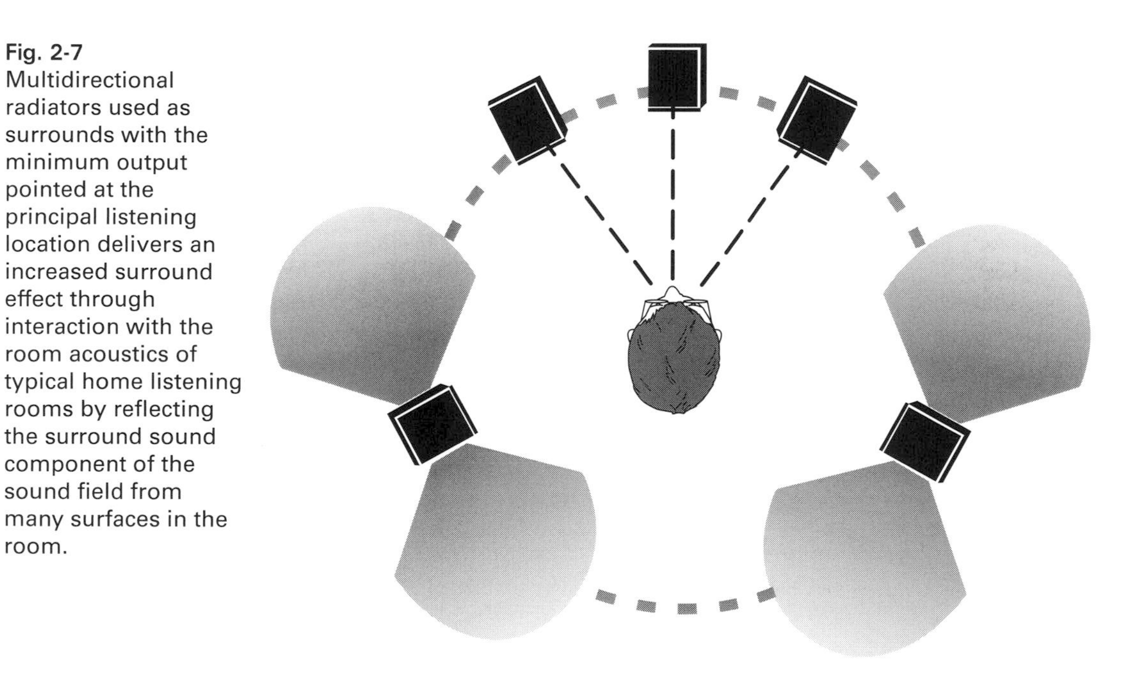

Fig. 2-7 Multidirectional radiators used as surrounds with the minimum output pointed at the principal listening location delivers an increased surround effect through interaction with the room acoustics of typical home listening rooms by reflecting the surround sound component of the sound field from many surfaces in the room.

Square Array

This places left and right at ±45° and surrounds at ±135°. The use of the center channel is generally minimized by these producers in their

work. The surround loudspeaker angle of ±110° for the AES/ITU setup was based on research that showed it to be the best trade-off between envelopment (i.e., best at ±90° when only 2 channels are available for surround) and rear quadrant imaging (which is better at ±135° than ±110°). A square array was thoroughly studied during the quad era as a means of producing sound all around, and information about the studies appears in Chapter 6. Nevertheless, it is true that increasing the surround angle from ±110° to ±135° improves rear phantom imaging, at the expense of envelopment.

One rationale given for the square array is the construction of four "sound fields," complete with phantom imaging capability, in each of the four quadrants front, back, left, and right. This thought does not consider the fact that human hearing is very different on the sides than in the front and back, due to the fact that our two ears are on the two sides of our heads. For instance, there is a strongly different frequency response in the ear canal for left front and left surround loudspeakers as we face forward, even if the loudspeakers are perfectly matched and the room acoustics are completely symmetrical. For best imaging all around, Günther Theile has shown that a hexagon of symmetrically spaced loudspeakers, located at ±30°, ±90°, and ±150°, would work well. So the current 5.1-channel system may be seen as somewhat compromised in the ability to produce sound images from all around. After all is said and done, the 5.1-channel system was developed for use in accompanying a picture, where there was a premium placed on frontal sound images and surround sound envelopment, not on producing sound images all round.

Close-Field Monitoring

An idea that sounds immediately plausible is "near-field monitoring," perhaps better termed close-field monitoring. The idea is that small loudspeakers, located close to the listener, put the listener in a sound field where direct sound predominates over reflected sound and reverberation. Thus such loudspeakers and positioning are allegedly affected less by the room acoustics than are conventional loudspeakers at a greater distance, and have special qualities to offer, such as flatter response (Fig. 2-8).

Unfortunately, in real-world situations, most of the advantages of the theory of near-field listening are unavailable, so the term close-field monitoring is probably more apt. The loudspeakers used are small compared to the range of wavelengths being radiated, and thus broadly radiate sound. The broadly radiating sound interacts with nearby surfaces strongly. For instance, "near-field" monitors on top of a console meter

Fig. 2-8 So called "near-field" monitoring suffers from the same problem as the elevated monitor, splashing excessive sound off the control surface. Moving the speaker to behind the console can be an improvement.

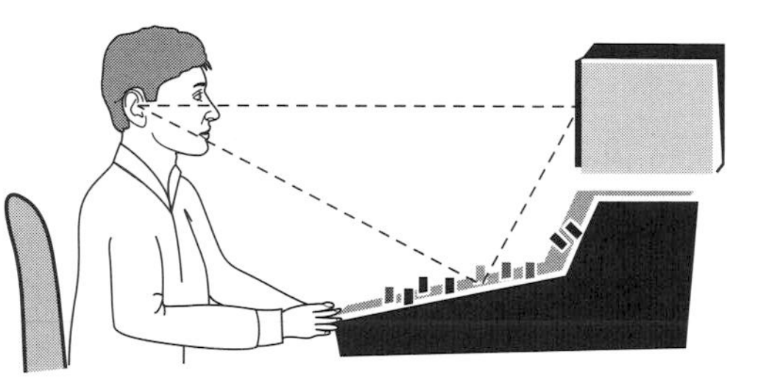

Near-field monitor also splashes sound off the control surface

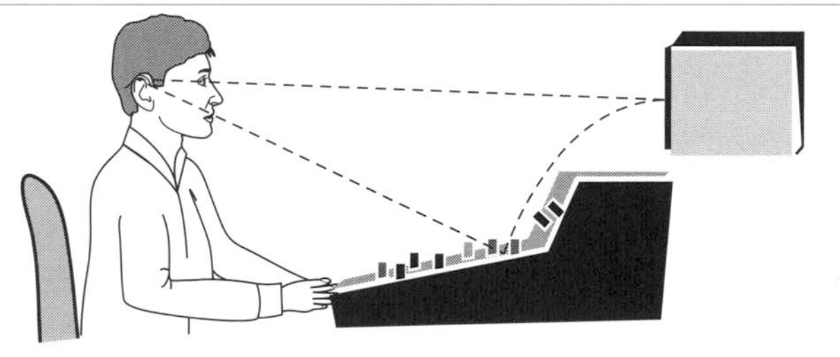

Loudspeaker mounted behind console uses it as a barrier, though sound diffracts over this console and still reflects, it is at a lower level than when mounted on the top

bridge reflect sound strongly off the console's operating surface, at levels well above audibility. Also, the console surfaces act to extend the baffle face of the monitor loudspeaker placed on top of the meter bridge, and this changes the mid-bass response. At mid-bass frequencies, the idea that the near-field monitor gets us out of trouble from interaction with the room is wrong. There is still just one transfer function (frequency response) associated with one source location (the loudspeaker) and one receiver location (the listener's head). The effects of standing waves occur at the speed of sound, which produces effects quickly in small rooms, so the theory that we are out of trouble because the speaker is close to the listener is wrong.

Another difficulty is with the crossover region between woofer and tweeter. When used at close spacing to the listener, the exact listening position can become highly critical in many designs, as the crossover strongly affects the output of the loudspeaker spatially. Only a 6 in. move on the part of the listener located a few feet away can make a dramatic shift in the monitor's direct sound frequency response, and be audible.

The close-field monitor arose as a replacement for the simple, cheap loudspeakers, usually located on top of the console that provided a "real-world" check on the large and professional built-in control room loudspeakers. In multichannel monitoring, the same theories could lead to having two systems, one huge and powerful, and the other close up and with lesser range and level capacity. This would be a mistake:

- The built-in type of control room monitors are often too high, which splashes sound off the console at levels above audibility.
- Close-field monitors suffer from the problems discussed above. They may not play loudly enough to operate at reference levels.

The best system is probably one somewhere in between the two extremes, that can play loudly enough without audible distortion to achieve reference level calibration and adequate headroom to handle the headroom on the source medium, with smooth and wide frequency range response, and otherwise meets the requirements shown in Table 2-2.

Time Adjustment of the Loudspeaker Feeds

Some high-end home controllers provide a function that is very useful when loudspeakers cannot be set up perfectly. They have adjustable time controls for each of the channels so that, for example, if the center loudspeaker has to be in line with the left and right ones for mounting reasons, then the center can be delayed by a small amount to place the loudspeaker effectively at the same distance from the listener as left and right, despite being physically closer.

The main advantage of getting the timing matched among the channels has to do with the phantom images that lie between the channels. If you pan a sound precisely halfway between left and center, and if the center loudspeaker is closer to you than the left one, you will hear the sound closer to the center than in the chosen panned position midway between left and center. The stereo sound field seems to flatten out around the center; that is, as the sound is just panned off left towards center, it will snap rather quickly to center then stay there as the pan continues through center until it is panned nearly to the right extreme, when it will snap to the right loudspeaker. The reason for this is the precedence effect. The earlier arriving center sound has precedence over the later left and right loudspeakers. This may be one reason that some music producers have problems with the center channel. Setting the timing correctly solves this problem.

Sound travels 1128 ft/s at room temperature and sea level. If the center is 1 ft closer to you than left and right, the center time will be 1.1 ms early. Some controllers allow you to delay center in 1 ms steps, and

1 ms is sufficiently close to 1.1 ms to be effective. A similar circumstance occurs with the surround loudspeakers, which are often farther away than the fronts, in many practical situations. Although less critical than for imaging across the front, setting delay on the fronts to match up to the surrounds, and applying the same principle to the subwoofer, can be useful.

The amounts of such delay are much smaller typically than the amounts necessary to have an effect on lip sync for dialog, for which errors of 20 ms are visible for the most sensitive listener/viewers. Note, however, that video pictures are often delayed through signal processing of the video, without a corresponding delay applied to the audio. It is best if audio-video timing can be kept to within 20 ms. Note that motion pictures are deliberately printed so that the sound is emitted by the screen speakers one frame (42 ms) early, and the sound is actually in sync 47 ft from the screen, a typical viewing distance.

Low-Frequency Enhancement—The 0.1 Channel

The 0.1 or LFE channel is completely different from any other sound channel that has ever existed before. It provides for more low-frequency headroom than on traditional media, just at those frequencies where the ear is less sensitive to absolute level, but more sensitive to changes in level. The production and monitoring problems associated with this channel potentially include the end user not hearing some low-frequency content at all, up through giving him so much bass that his subwoofers blow up.

With adoption by film, television, and digital video media already, the 0.1 channel has come into prominence over the past few years. With music-only formats, how will it affect them? Just what is the "0.1" channel? How do we get just part of a channel? Where did it come from, and where is it going? And most importantly of all, how does a professional apply the standards that have been established to their application, from film through television to music and games?

Film Roots

The beginnings of the idea came up in 1977 from the requirements of Gary Kurtz, producer of a then little-known film called *Star Wars*. At the time, it seemed likely that there would be inadequate low-frequency headroom in the three front channels in theaters to produce the amount of bass that seemed right for a war in outer space (waged in a vacuum!). A problem grew out of the fact that by the middle 1970s, multichannel film production used only three front channels—left, center, and right (two "extra" loudspeaker channels called left extra and right extra, in between left and center, and center and right, were used in the 1950s Todd AO and Cinerama formats). The loudspeakers employed in most theaters were various models of Altec-Lansing Voice of the Theater. A problem with these loudspeakers was that although they use horn-loaded operation across much of the woofer's operating range for high efficiency, and thus high-level capability, below about 80 Hz the small size of the short horn became ineffective, and the speaker reverted to "bass reflex" operation with lower efficiency. In addition, the stiff-suspension, short-excursion drivers could not produce much level at lower frequencies, and the "wing walls" surrounding the loudspeakers that supported the bass had often been removed in theaters. For this combination of reasons, both low frequency response and headroom were quite limited.

Ioan Allen and Steve Katz of Dolby Labs knew that many older theaters were still equipped with five-front channels, left over from the original 70mm format. So, their idea was to put the "unused" channels back into service just to carry added low-frequency content, something that would help distinguish the 70 mm theater experience from the more ordinary one expected of a 35 mm release at the time. Gary Kurtz asked for a reel of *Capricorn1* to be prepared in several formats and played at the Academy theater. The format using left extra and right extra loudspeakers for added low-frequency level capability won. Thus was invented the "Baby Boom" channel, as it was affectionately named, born of necessity. Just 6 months after *Star Wars, Close Encounters of the Third Kind* was the first picture to use dedicated subwoofers, installed just for the purpose of playing the Baby Boom channel.

Headroom on the Medium

Another idea emerged along with that of using more speakers for low frequencies. The headroom of the magnetic oxide stripe on the 70mm film used in 1977 was the same as it was before 1960 because the last striping plant that remained was still putting on the same oxide! Since the headroom on the print was limited to 1960s performance by the old oxide, it was easy to turn down the level recorded on the film by 10 dB, and turn the gain back up in the cinema processor by 10 dB, thus improving low-frequency headroom on the film in the boom channel, although also suffering from a greater susceptibility to hum, but that could be looked after. At first, a 250 Hz low-pass filter stripped off the higher frequencies, so hiss was no problem. Within a few films, the frequency of the low-pass filter on the boom channel was lowered to 125 Hz. This was done so that if speech was applied to the boom channel, it would not get thick-sounding, because the usual boom channel content was just the sum of the other channels low-pass filtered. Even for *Star Wars*, special content for the boom channel was being recorded that did not appear in the main channels, so a whole new expressive range was exercised for high-level low-frequency sounds. In fact at Lucasfilm about 1982 Ben Burtt asked me for a "boom sting" button, a button that when pressed just upped the gain temporarily by 6 dB—faster than using a fader.

The idea of having more low-frequency than mid-range headroom proved to be a propitious one. Modern day psychoacoustics tells us that human listeners need more low-frequency sound pressure level to sound equally as loud as a given mid-range level. This is known from the equal loudness contours of hearing, often called the Fletcher–Munson curves, but modernized and published as ISO 226:2003. (The 1930s Fletcher–Munson curves show the equal loudness curve corresponding to 100 dB sound pressure level at 1 kHz to be nearly flat in the bass, but all later experimenters find you need more bass to keep up with the loudness of the mid-range.) Observing these curves demonstrates the need for more low-frequency level capability than mid-range capability, so that a system will sound as though it overloads at the same perceived level versus frequency (Fig. 2-9).

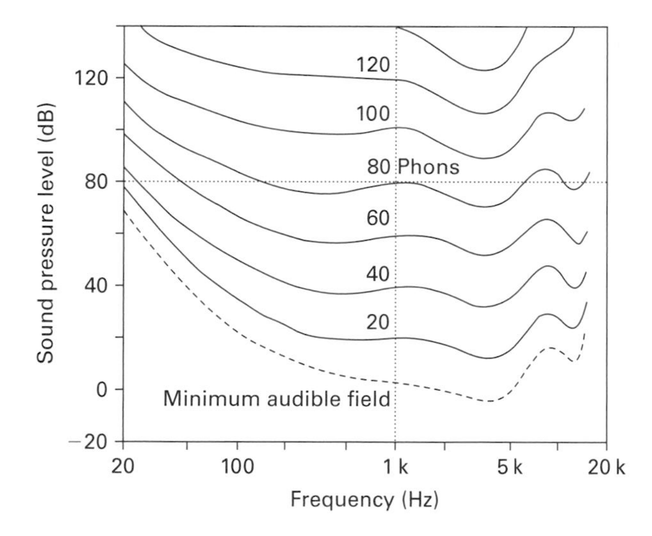

Fig. 2-9 Equal loudness contours of hearing, sound pressure level in dB re 20 μN/m² (0 dB SPL is about the threshold of hearing) versus frequency in Hz, from ISO 226. These curves show that at no level is the sensation of loudness flat with frequency; more energy is required in the bass to sound equally as loud as the mid-range.

By the way, I found out about this in a completely practical way. Jabba the Hutt's voice in *Jedi* started out with a talker with a very deep voice (a then-marketing guy from Dolby, Scott Schumann) recorded on a directional microphone used up close, thus boosting the bass through the proximity effect.

Then the voice was processed by having its frequency lowered by a pitch shifter, and a sub-harmonic synthesizer was used to provide frequencies at one-half the speech fundamental, adding content between 25 and 50 Hz corresponding to the frequencies in the speech between 50 and 100 Hz. All of these techniques produce ever greater levels of low bass. When I looked at the finished optical sound track of a Jabba conversation, I found that he used up virtually 100% of the available area of the track. Yet Jabba's voice is not particularly loud. Thus, really low bass program content can "eat up" the available headroom of a system quickly. This lesson surely is known to engineers who mix for CDs.

Digital Film Sound Enters the Picture

When digital sound on film came along as the first digital multichannel format, laid out in 1987 and commercially important by 1993, the 0.1 channel was added to the other five in order to have a low-frequency channel capable of greater headroom than the main channels, following the theory developed for 70 mm film. Note that the five main channels have full frequency range capability extending downwards into the infrasonic region on the medium, whether film, DVD, Digital Television, or any other, so the 0.1 channel provides an added headroom capability, and does not detract from the possibility of "stereo bass."

I proposed the name "5.1" in a Society of Motion Picture and Television Engineers (SMPTE) meeting of a short-lived subcommittee called Digital Sound on Film in October 1987. The question as to the number of channels was going around the room, and answers were heard from four through eight. When I said 5.1, they all looked at me as though I were crazy. What I was getting at was that this added channel worth of information, with great benefits for low-frequency headroom, only took up a narrow fraction of one of the main channels. In fact, the amount is really only 1/200th of a channel, because the sample rate can be made 1/200th of the main channel's sample rate (240 Hz sampling, and 120 Hz bandwidth for a 48 kHz sampled system), but the number 5.005 just didn't quite roll off the tongue the way 5.1 did. Call the name "marketplace rounding error," if you will.

Bass Management or Redirection

In theater practice, routing of the signals from the source medium into the loudspeakers is simple. The main loudspeaker channels are "full range" and, thus, each of the five main channels on the medium is routed to the respective loudspeakers, and the 0.1 channel is routed to one or more subwoofers. This seemed like such an elegant system, but it did have a flaw. When I played Laser Discs at home, made from film masters, I noticed that my wide-range home stereo system went lower in the bass than the main channel systems we used in dubbing. So what constitutes a "full-range" channel? Is it essentially flat to 40 Hz and rolled off steeply below there like modern direct-radiator, vented-box main channel theater systems installed in large, flat baffle walls? Or is it the 25 Hz range or below of the best home systems? This was disturbing, because my home system revealed rumbles that were inaudible, or barely audible when you knew where they were, on the dubbing stage system, even though it had the most bass extension of any such professional system.

During this time, I decided to use a technique from satellite–subwoofer home systems, and sum the lowest bass (below 40 Hz) from the five main channels and send it to the subwoofer, along with the 0.1-channel content. Thus, the subwoofer could do double duty, extending the main channels downwards in frequency, as well as adding the extra 0.1-channel content with high potential headroom. All this so the re-recording mixers could hear all of the bass content being recorded on the master.

This sounds like a small matter at first thought. Does a difference in bass extension from 40 to 25 Hz really count? As it turns out, we have to look back at psychoacoustics, and there we find an answer for why these differences are more audible than expected from just the numbers. Human hearing is more sensitive to changes in the bass than changes in the mid-range.

The equal loudness contour curves are not only rising in the bass, but they are converging as well. This means that a level change of the lowest frequencies is magnified, compared to the same change at mid-range frequencies. Think of it this way: as you go up 10 dB at 1 kHz, you cross 10 dB worth of equal loudness contours, but as you go up 10 dB at 25 Hz, you cross more of the contour lines—the bass change counts more than the mid-range one. This does not perhaps correspond to everyday experience of operating equalizers, because most signal sources contain more mid-range content compared to very low frequencies, so this point may seem counterintuitive, but it is nonetheless true. So, since extending the bass downwards in frequency by just 15 Hz (from 40 to 25 Hz) changes the level at the lower frequency by a considerable amount, the difference is audibly great.

Digital Television Comes Along

When the 5.1-channel system was adopted for Digital Television, a question arose from the set manufacturers: What to do with the 0.1 channel?

Surely not all televisions were going to be equipped with subwoofers, yet it was important that the most sophisticated home theaters have the channel available to them. After all, it was going to be available from DVD, so why not from Digital Television? A change in the definition of how the 0.1 channel is thought of occurred. The name "Low-Frequency Enhancement" (LFE) was chosen to describe what had been called the Baby Boom or 0.1 channel up until that time. The LFE name was meant to alert program providers that reproduction of the 0.1-channel content over television is optional on the part of the end user set. In theatrical release, having digital playback in a theater ensures that there will be a subwoofer present to reproduce the channel. Since this condition is not necessarily true for television, the nature of the program content that is to be recorded in the channel changes. To quote from ATSC Standard A/54, "Decoding of the LFE channel is receiver optional. The LFE channel provides non-essential low-frequency effects enhancement, but at levels up to 10 dB higher than the other audio channels. Reproduction of this channel is not essential to enjoyment of the program, and can be perilous if the reproduction equipment cannot handle high levels of low-frequency sound energy. Typical receivers may thus only decode and provide five audio channels from the selected main audio service, not six (counting the 0.1 as one)."

This leads to a most important recommendation: *Do not record essential story-telling sound content only in the LFE channel for digital television.* For existing films to be transferred to DTV, an examination should be made of the content of LFE to be certain that the essential story-telling elements are present in some combination of the five main channels, or else it may be lost to many viewers. If a particular recording exists only in the LFE channel, say the sound effect of the Batmobile, then the master needs remixing for television release. Such a remix has been called a "Consumer 5.1" mix.

Bass management brings with it an overhead on the headroom capacity of subwoofers and associated amplifiers. Since bass management sums together 5 channels with the LFE channel at +10 dB gain (which is what gives the LFE channel 10 dB more headroom), the sum can reach rather surprisingly high values. For instance, if a room is calibrated for 85 dB (C-weighted, slow) with −20 dBFS noise (see level calibration at the end of this chapter), then the subwoofer should be able to produce a sound pressure level of 121 dB SPL in its passband! That is because 5 channels plus the LFE, with the same signal in phase on all the channels, can add to such a high value. Film mixers have understood for a long time that to produce the highest level of bass, the signal is put in phase in all of the channels, and this is what they sometimes do, significantly "raising the bar" for bass managed systems.

Home Reproduction

When it comes to home theater, a version of the satellite–subwoofer system is in widespread use, like what we had been doing in dubbing stages for years, but with higher crossover frequencies. Called bass management, this is a system of high-pass filtering the signals to the five main channels, in parallel with summing the 5 channels together and low-pass filtering the sum to send to the subwoofer. Here, psychoacoustics is useful too, because it has been shown that the very lowest frequencies have minimally audible stereophonic effect, and thus may be reproduced monophonically with little trouble. Work on this was reported in two important papers: "Perceptibility of Direction and Time Delay Errors in Subwoofer Reproduction," by Juhani Borenius, AES Preprint 2290, and "Loudspeaker Reproduction: Study on the Subwoofer Concept," by Christoph Kügler and Günther Theile, AES Preprint 3335.

Many people call the LFE channel the "subwoofer channel." This idea is a carry over from cinema practice, where each channel on the medium gets sent to its associated loudspeaker. Surely many others outside of cinema practice are doing the same thing. They run the risk that their "full-range" main channel loudspeakers are not reproducing the very lowest frequencies, and home theater listeners using a bass management system may hear lower frequencies than the producer! This could include air conditioning rumble, thumps from the conductor stomping on the podium, nearby subways, and analog tape punch-in thumps, to mention just a few of many other undesired noises. Of course desirable low-frequency sound may also be lost in monitoring.

In fact, the LFE channel is the space on the medium for the 0.1 channel, whereas the "subwoofer channel" is the output of the bass management process, after the lowest frequencies from the five main channels have been summed with the 0.1-channel content.

0.1 for Music?

The LFE channel came along with the introduction of AC-3 and Digital Theater System (DTS) to the Laser Disc and DVD media for the purpose of reproducing the channel that had been prepared for the theater at home. Naturally then, it seems logical to supply the function for audio-only media, like multichannel disc formats. While many may question the utility of the added low-frequency headroom for music, multichannel music producers are already using the channel. Among its advantages are not just the added LF headroom, but the accompanying decrease in intermodulation distortion of the main channel loudspeakers when handling large amounts of low bass. If the bass required to sound loud were to be put into the main channels, it would cause such intermodulation, but in a separate channel, it cannot. Whether this is audible or not is certainly debatable, but it is not debatable that having a separate channel reproduced by a subwoofer eliminates the possibility of intermodulation (except in the air of the playback room at really high levels!).

What is most important to the use of LFE for music is the understanding that standards exist for the bandwidth and level of the LFE channel compared to the main channels. Since the bandwidth is controlled by the media encoders, getting the monitor level for the 0.1 channel right is the bottom line.

First, however, it is important to know that what you are monitoring has the correct bandwidth. If you record a bass drum to the LFE channel of a digital multitrack, then play it back in your studio to a subwoofer, you have made a mistake. The problem is that the only bandwidth limiting being done is the high-frequency limit of your subwoofer. You may be very surprised to find that after the tape is mastered, the bass drum has lost all of its "thwack," because the bandwidth limitation of the LFE channel has come into play. Correct bass management in the studio will allow you to hear what the format encoder is going to do to the LFE channel, and you can mix the higher frequencies of the bass drum into the main channels, as well as its fundamentals into the LFE channel, for best reproduction.

Note that all systems employing the 5.1- or 7.1-channel configurations, whether they are on film or disc, or intended for broadcast, and whether coded by linear PCM, AC-3, DTS, or MPEG, all have two vital specifications that are the same: the sample rate of the LFE channel is 240 Hz for 48 kHz sampled systems leading to practically 120 Hz bandwidth (and proportionately lower for 44.1 kHz systems), and the intended playback level is +10 dB of "in-band gain" compared to the main channels (Fig. 2-10).

"In-band gain" means that the level in each 1/3-octave band in the main operating range of the subwoofer is 10 dB above the level of each of the 1/3-octave bands of one of the main channels, averaged across its main frequency range. This does not mean that the level measured with a sound level meter will measure 10 dB higher, when the LFE channel is compared to a main channel. The reason for this apparent anomaly is that the bandwidth of the main channel is much wider than that of the LFE channel, which leads to the difference—there's more overall energy in a wider bandwidth signal. In an emergency, you could set the LFE level with a sound level meter. It will not read 10 dB above the level of broadband pink noise for the reason explained, but instead about 4 dB, when measuring with a C-weighting characteristic available even on the simple Radio Shack sound level meter. There are many possible sources of error using just a sound level meter, so this is not a recommended practice, but may have to do in a pinch.

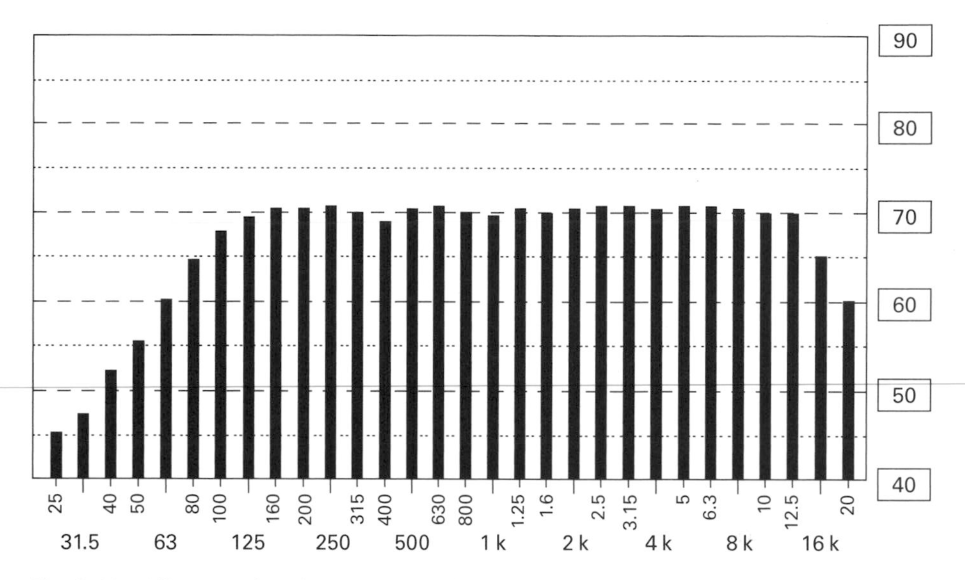

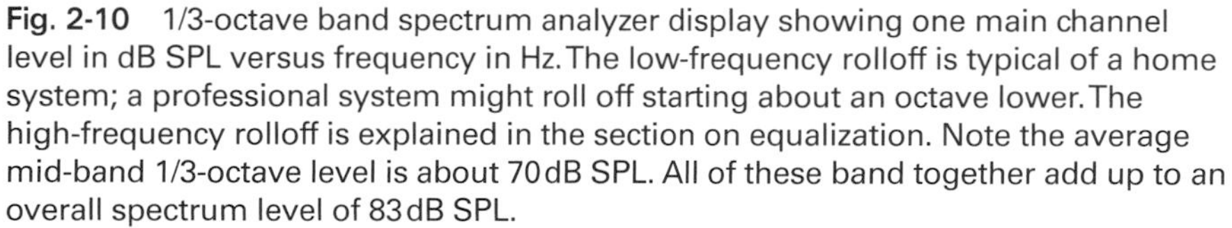

Fig. 2-10 1/3-octave band spectrum analyzer display showing one main channel level in dB SPL versus frequency in Hz. The low-frequency rolloff is typical of a home system; a professional system might roll off starting about an octave lower. The high-frequency rolloff is explained in the section on equalization. Note the average mid-band 1/3-octave level is about 70 dB SPL. All of these band together add up to an overall spectrum level of 83 dB SPL.

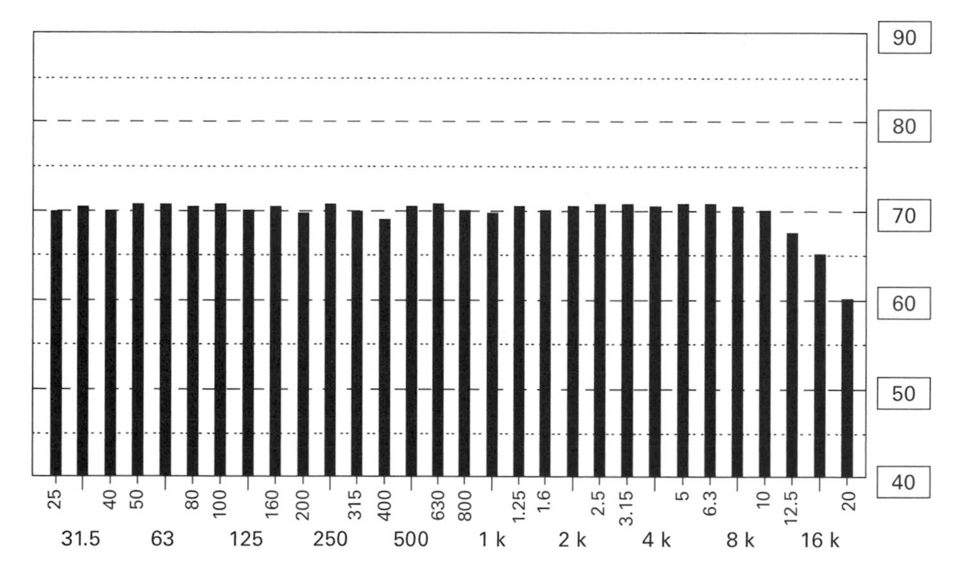

Fig. 2-11 1/3-octave band spectrum analyzer display showing level in dB SPL versus frequency in Hz of a main channel spliced to a subwoofer. This is one of the jobs of bass management—to extend the low-frequency limit on each of the main channels by applying the correct signal to one or more subwoofers.

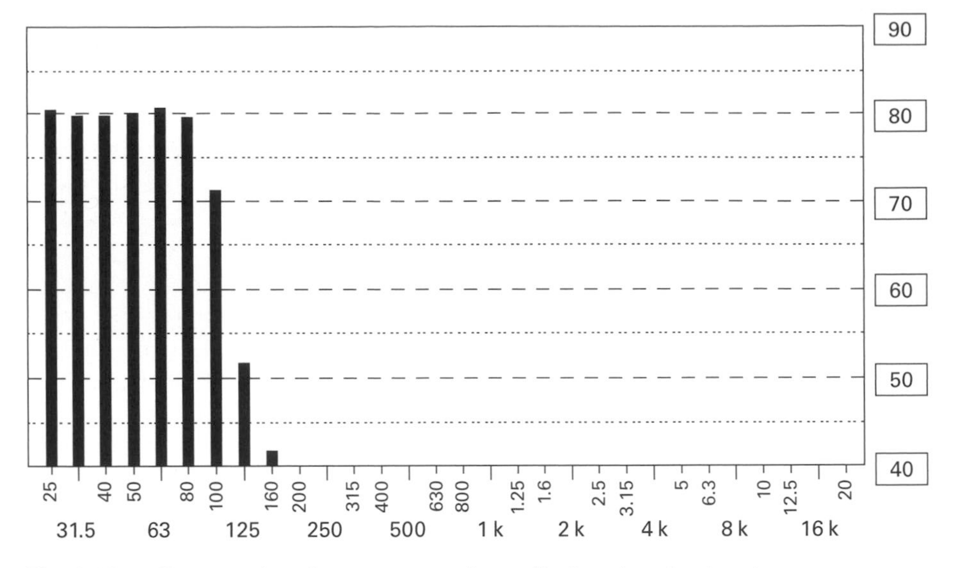

Fig. 2-12 1/3-octave band spectrum analyzer display showing level versus frequency of a properly aligned LFE channel playing over the same subwoofer as used above. The level of pink noise on the medium is the same as for Fig. 2-11, but the reproduction level is +10 dB of in-band gain.

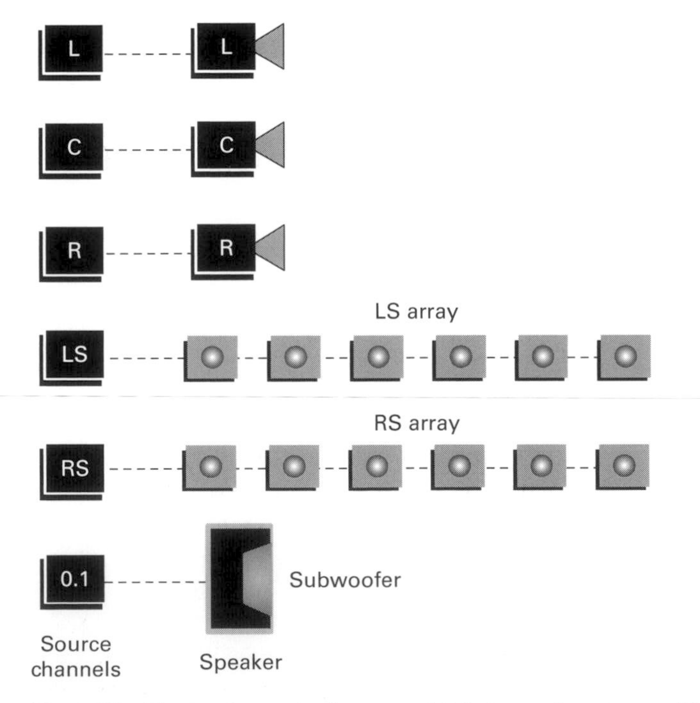

Fig. 2-13 Block schematic diagram of 5.1-channel systems without bass management. This is typical of all motion picture theaters and most film dubbing stages and television mixing rooms. Although the main channels are "wide range," they typically roll off below 40 Hz, so the very lowest frequencies are attenuated in the main channels. This can lead to not hearing certain problems, covered in the text.

An anti-aliasing, low-pass filter is included in media encoders, such as those by Dolby and DTS, in the LFE channel. If you were to listen in the studio to a non-band-limited LFE source over many subwoofer models, you would hear program content out to perhaps 1–2 kHz, which would then subsequently be filtered out by the media encoder. This means you would hear greater subwoofer bandwidth from your source channels than after encoding. Thus it is important in the studio to use a low-pass filter in monitoring the LFE channel when a media encoder is not in use. Characteristics of this filter are a bandwidth of 110 Hz, and a very steep slope.

Typical specs for the filters in pro-audio might be: (1) 50 Hz 2-pole Butterworth (12 dB/octave) high pass in each main channel; (2) two 50 Hz 2-pole Butterworth (12 dB/octave each) low-pass filters in series in the summed subwoofer path; and (3) 110 Hz steep anti-aliasing filter in the LFE feed. Summing the electrical filters and the response of the main channel speakers and subwoofer produces a fourth-order Linkwitz-Riley acoustic response. The slopes (outside the high-pass/low-pass symbols) and frequencies (given inside the symbols) represent typical professional system use (Fig. 2-14).

Typical specs for the filters for high-quality home theater are: (1) 80 Hz 2-pole Butterworth high pass in each main channel, (2) two 80 Hz 2-pole Butterworth low-pass filters in series in the summed subwoofer path; and (3) anti-aliasing filter for the LFE channel, built into the format decoder.

The Bottom Line

- The 0.1 channel, called LFE, is provided for more headroom below 120 Hz, where the ear is less sensitive and "needs" more SPL to sound equally loud as mid-range sound.
- LFE is a monophonic channel. For stereo bass, use the five main channels.
- Bass management in monitoring can be used to reproduce both the very low-frequency content of the main channels, as well as the LFE channel, over one or more subwoofers.
- The LFE channel recorded reference level is −30 dBFS for masters using −20 dBFS reference on the main channels.
- When both are measured in 1/3-octave bands using pink noise at the same electrical level, the LFE channel 1/3-octave band SPL reference is 10 dB above the level of one main channel; the pink noise for the LFE channel is band limited to 120 Hz, and is wideband for the main channels. (See Figs. 2-11 and 2-12, comparing the in-band level in the subwoofer operating region.) Typically, LFE will measure about 4 dB above the SPL of one main channel playing

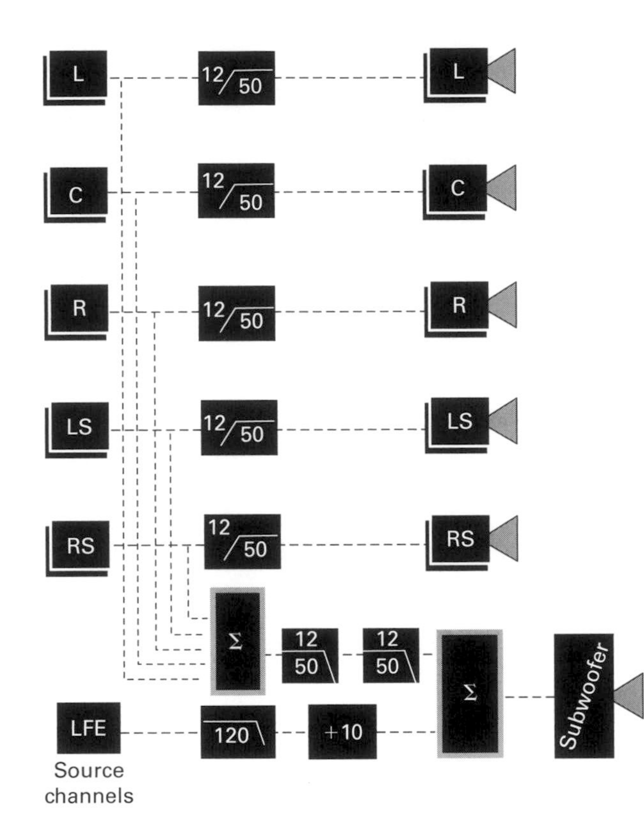

Fig. 2-14 Block diagram of a bass management system, with typical characteristics for pro-audio shown. High-pass filters in each of the main channels are complemented by a low-pass filter set in the subwoofer feed, considering the effects of the loudspeaker responses, so that each of the channels is extended downwards in frequency with a flat acoustical response. Note that this requires five matched bandwidth loudspeakers. The LFE channel is low-pass filtered with an anti-aliasing filter, which may be part of the encoding process or be simulated in monitoring, with its level summed into the bass extension of the main channels at +10 dB relative to one main channel.

pink noise on a C-weighted sound level meter. The LFE level is not +10 dB in overall sound pressure level compared to a main channel because its bandwidth is narrower.

Calibrating the Monitor System: Frequency Response

Equalizing monitor systems to a standard frequency response is key to making mixes that avoid defects. This is because the producer/engineer equalizes the program material to what sounds good to them (and even if not doing deliberate equalization, still chooses a microphone and position relative to the source that implies a particular frequency response), and a bass-shy monitor, for instance, will cause them to turn up the bass in the mix. This is all right only if all the listeners are listening to the same monitor system; if not, then monitor errors lessen the universality of the mix. Although room equalization has a bad name among some practitioners because their experience with it has been bad, that is because there has been bad equalization done in the past.

I did an experiment comparing three different equalization methods with an unequalized monitor. I used a high-quality contemporary monitor loudspeaker in a well-qualified listening room, multiple professional listeners, multiple pieces of program material, and double-blind, level-matched experimental conditions. The result was that all three equalization methods beat the unequalized condition for all the listeners on virtually all the program material. Among the three methods of equalizing, the differences were much smaller than between equalizing and not equalizing. You will not hear this from loudspeaker manufacturers typically, but it is nonetheless true.

Considerations in choosing a set of equalizers and method of setting them are:

- The equalizer should have sufficient resources to equalize the effects of rooms acoustics; this generally means it has many bands of equalization.
- The method of equalization should employ spatial consolidation. Measuring at just one point does not well represent even human listeners with two ears. Averaging or clustering the responses and nominating one best response based at several points generally leads to less severe, and better sounding, equalization.
- Equalizers that fix just the direct field, such as those digital equalizers that operate only in the first few milliseconds, seem to be less useful in practical situations than those that fix the longer-term or steady-state response.
- If a noise-like signal is the test source for equalizing, temporal (time) averaging is necessary to produce good results. For 1/3-octave band analysis, averaging for 20 seconds generally produces small enough deviations. Trying to average the bouncing digits of a real-time analyzer by eye produces large errors.

A Choice of Standardized Response

Film sound uses the standards ISO 2969 and SMPTE 202 for the target frequency response of the monitor system. Called the X curve for extended, wide-range response, this is a nationally and internationally recognized standard that has helped interchangeability of film program material throughout the world. The US standard includes the method of measurement along with the curve (Fig. 2-15).

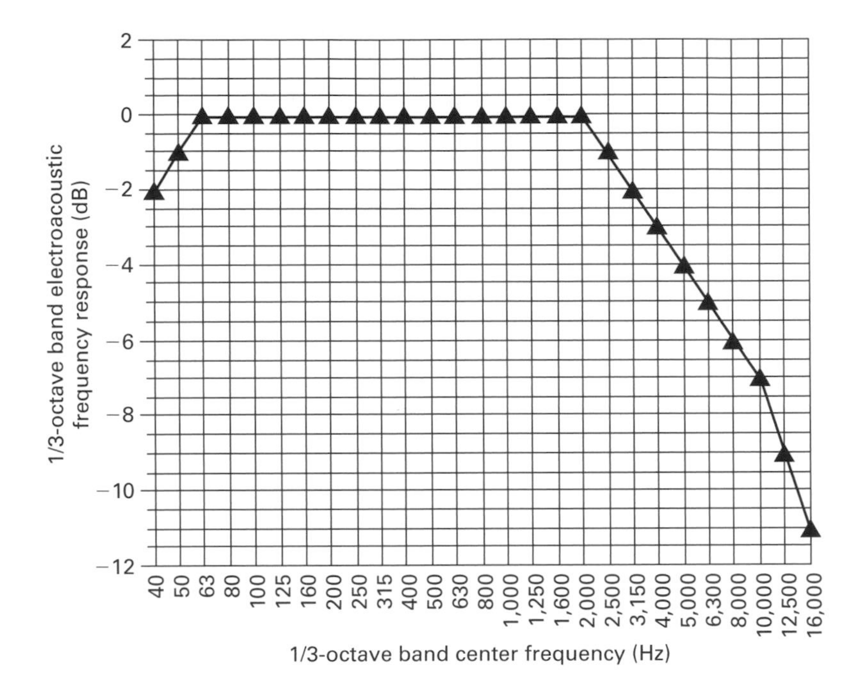

Fig. 2-15 The X curve of motion picture monitoring, to be measured spatially averaged in the listening area of the sound system with quasi-steady-state pink noise and low-diffraction (small) measurement microphones. The room volume must be at least 6,000 ft^3. The curve is additionally adjusted for various room volumes; see SMPTE 202.

Television and music have no such well-established standard. They tend to use a monitor loudspeaker that measures flat anechoically on axis, thus making the direct sound flat at the listener (so long as the loudspeaker is aimed at the listener, and neglecting air loss that is extremely small in conventional control rooms). Depending on the method of measurement, this may or may not appear flat when measured at the listening location, which also has the effects of discrete reflections and reverberation.

A complication in measuring loudspeakers in rooms is that the loudspeaker directivity changes with frequency, and so does the reverberation time. Generally, loudspeakers become more directional at high frequencies, and reverberation time falls. The combination of these two means that you may be listening in the reverberant-field dominated area at low and middle frequencies, but in the direct sound dominated area at high frequencies. Thus at high frequencies the direct sound is more important than the steady state. Measured with pink noise stimulus, correctly calibrated microphones, and spatial and time averaged spectrum analysis, the frequency response will not measure flat when it is actually correct. What is commonly found in control rooms is that the response is flat to between 6.3 and 12.5 kHz with a typical break frequency from flat of 10 kHz, and then rolls off at 6 dB/octave. Basically, if I know that a monitor loudspeaker is indeed flat in the first arrival sound (and I can measure this with a different measurement method), I do not boost high frequencies during equalization. In fact, most of the equalization that is done is between 50 and 400 Hz, where the effects of standing waves dominate in rooms.

For monitoring sound from film, there needs to be a translation between the X curve and normal control room monitoring, or the film program material will appear to be too bright. This is called re-equalization, and is a part of HomeTHX. When playing films in a studio using a nominally flat monitor response such as described above, addition of a high-frequency shelf of −4 dB at 10 kHz will make the sound better.

Calibrating the Monitor System: Level

Once the monitor system has been equalized, the gain must be set correctly for each channel in turn, and in a bass managed system the subwoofer level set to be correct to splice to the main channels and extend them downwards in frequency, with neither too little nor too much bass. If the bass management circuitry is correct, the in-band gain of LFE will then be the required +10 dB (Fig. 2.16).

There is a difference in level calibration of motion picture theaters and their corresponding dubbing stages on the one hand, and control rooms and home theaters on the other. In film work, each of the two surround channels is calibrated at 3 dB less than one of the screen channels; this is so the acoustical sum of the two surround channels adds up to equal one screen channel. In conventional control rooms and home theaters the calibration on each of the 5 channels is for equal level. An adjustment of the surround levels down by 3 dB is necessary in film transfers to home theater media, and at least one media encoder includes this level adjustment in its menus.

Proper level setting relies on setting the correct relationship between studio bus level and sound pressure level at the listening location. In any one studio, the following may be involved:

- a calibrated monitor level control setting on the console;
- any console monitor output level trims;
- room equalizer input and/or output gain controls;
- power amplifier gain controls;
- loudspeaker sensitivity; or
- in the case of powered loudspeakers, their own level controls.

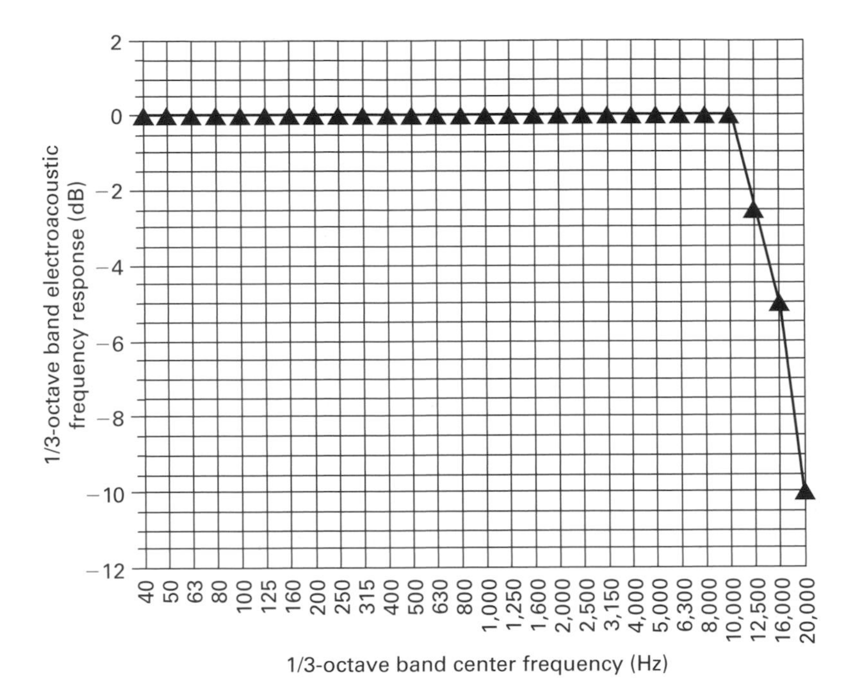

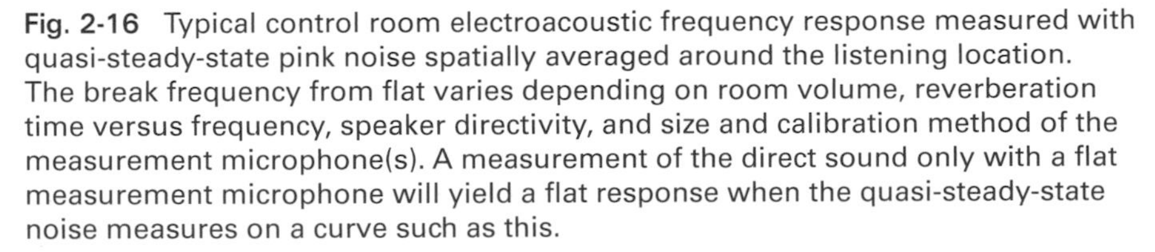

Fig. 2-16 Typical control room electroacoustic frequency response measured with quasi-steady-state pink noise spatially averaged around the listening location. The break frequency from flat varies depending on room volume, reverberation time versus frequency, speaker directivity, and size and calibration method of the measurement microphone(s). A measurement of the direct sound only with a flat measurement microphone will yield a flat response when the quasi-steady-state noise measures on a curve such as this.

You must find a combination of these controls that provides adequate headroom to handle all of the signals on the medium, and maintains a large signal-to-noise ratio. The test tape available from, Martinsound[4] is an aid to adjusting and testing the dynamic range of your monitor system. The resulting work is called "gain staging," which consists of optimizing the headroom versus noise of each piece of equipment in the chain. High-level "boinks" are provided on the test tape that check headroom for each channel across frequency. By systematic use of these test signals, problems in gain staging may be overcome. As part of gain staging, one level control per loudspeaker must be adjusted for reference level setting. Some typical monitor level settings are given in Table 2-4.

The best test signal for setting level electrically is a sine wave, because a sine wave causes steady, unequivocal readings. In acoustical work, however, a sine wave does not work well. Try listening to a 1 kHz sine wave tone while moving your head around. In most environments

[4]http://www.martinsound.com/pd_mch.htm

Table 2-4 Reference Levels for Monitoring

Type of program	*SPL* for −20 dBFSave*
Film	85
Video	78
Music	78–93

*Sound pressure level in dB re 20μN/m². SMPTE RP 200–2002 specifies an average responding, rms calibrated detector with pink noise at −20 dBFS. The sound level meter is to be used with C-weighting and the "slow" (1 s) detector time constant.

you will find great level changes with head movements, because the standing waves affect a single frequency tone dramatically. Thus, noise signals are usually used for level setting acoustically, because they contain many frequencies simultaneously. Pink noise is noise that has been equalized to have equal energy in each octave of the audio spectrum, and sounds neutral to listeners; therefore, it is the usual source used for level setting.

A problem creeps in with the use of noise signals; however, they show a strong and time-varying difference between their rms level (more or less the average) and peak level. The difference can be more than 10 dB. So which level is right? The answer depends on what you are doing. For level setting, we use the average or rms level of the noise, both in the electrical and in the acoustical domains. Peak meters, therefore, are not useful for this type of level setting, as they will read (variably) about 10 dB too high. The best we can do is to use a sine wave of the correct average level to set the console and recorder meters, then use the same level of noise, and set the monitor channel gain, 1 channel at a time, so that the measured SPL at the listening location reads the standard in Table 2-4.

An improvement on wideband pink noise is to use filtered pink noise with a two-octave bandwidth from 500 Hz to 2 kHz. This noise avoids problems at low frequencies due to standing waves, and at high frequencies due to the calibration and aiming of the measurement microphone. Test materials using sine wave tones to calibrate meters, and noise at the same rms level to calibrate monitors, are available from the author's web site, www.tmhlabs.com.

On these test materials a reference sine wave tone is recorded at −20 dBFSrms to set the console output level on the meters, and filtered pink noise over the band from 500 Hz to 2 kHz is recorded at a level of −20 dBFSrms to set the acoustical level of the monitor with a sound level meter. Once electrical level is set by use of the sine wave tone and console meters, the meters may be safely ignored, as they may read from 1 dB low (true VU meters, and so-called loudness meters that use an average responding detector instead of an rms one) to more than 10 dB high (various kinds of peak meters). It is customary to turn up the console fader by the 1 dB to make the average level of the noise read 0 VU, but be careful that what you are looking at is actually a VU meter to the IEEE standard. It should read about 1 dB low average on the noise compared to the tone.

For theatrical feature work this level of noise is adjusted to 85 dB on a sound level meter set to C-weighting and slow reading located at the primary listening position. For television use on entertainment programming mixed in Hollywood, reference level ranges from about 78 to 83 dB SPL. For music use, there is no standard, but typical users range from 78 up to 93 dB. TMH 10.2-channel systems are calibrated to 85 dB SPL for −20 dBFSrms.

For each channel in turn, adjust the power amplifier gain controls (or monitor equalization level controls) for the reference sound pressure level at the listening location. A Radio Shack sound level meter is the standard of the film industry and is a simple and cheap method to do this. The less expensive analog Radio Shack meter is preferred to the digital one because it can be read to less than 1 dB resolution, whereas the digital meter only shows 1 dB increments.

3 Multichannel Microphone Technique

Tips from This Chapter

- Various stereo methods may be extended to multichannel use with a variety of techniques. The easiest is pan-potted stereo, wherein each mono source is panned to a position in the sound field, but other methods such as spaced omnis and coincident techniques may also be expanded to multichannel use, with some restrictions and problems given in the text.
- The basic surround sound decision breaks down into two ways of doing things: the direct/ambient approach, and the sources-all-round approach. The direct/ambient method is much like most attentive listening, with sources generally in front of one, and reflections and reverberation of natural acoustic spaces coming from multiple directions. The sources-all-round approach may be viewed as more involving, or as disturbing, depending on the listener, and may expand the vocabulary for audio to affect composition and the sound art.
- Reverberation may be recorded spatially with multiple microphones, using for instance the rejection side of a cardioid microphone aimed at the source to pick up a greater proportion of room sound than direct sound.
- Spot miking is enhanced with digital time delay of the spot mike channels compared to the main channels.
- Microphone setups for multichannel include use of standard microphones in particular setup combinations, and microphone systems designed as a whole for multichannel sound.
- Complex real-world production especially accompanying a picture may "overlay" several of the multichannel recording techniques for the benefit of each type of material within the overall program.

- A method is given for recording the elements needed for 2- and 5.1-channel releases on one 8-track format recording, including provision for bit splitting so that 20-bit recording is possible on a DTRS format (DA-98) machine for instance.

Introduction

There are many texts that cover stereophonic microphone technique and a useful anthology on the topic is *Stereophonic Techniques*, edited by John Eargle and published by the Audio Engineering Society. This chapter assumes a basic knowledge of microphones, such as the fundamental pressure (omni) and pressure-gradient (Figure-8) transducers and their combination into cardioid and other polar patterns, and so forth. If you need information on these topics, see one of the books that include information about microphones.[1]

Using the broadest definition of stereo, the several basic stereo microphone techniques are:

- Multiple microphones usually closely spaced to the sources, pan potted into position, usually called "pan pot stereo."
- Spaced microphones, usually omnis, spread laterally across a source, called "spaced omnis" and represented by the Decca Tree among others, and often aided by "spot mikes" on individual sources.
- Coincident or near-coincident directional microphones; including crossed Figure-8 (Blumlein), X-Y, M-S, ORTF, Ambisonics, and others.
- Barrier-type techniques including Faulkner, Schoeps sphere, and Holophone.
- Dummy head binaural.
- Techniques that include combinations of elements of these various methods, such as film sound that may employ one technique for one layer of a sound track called a stem, and various other techniques for other stems.

In addition, there are some new methods that are designed for specifically for surround sound. These include:

- Double M-S.
- Fukada Array, similar to the Decca Tree only uses spaced cardioids.
- Hamasaki square, spaced bidirectional mikes arranged in a 2–3 m sided square with their nulls pointed at the main source and their positive polarity sides pointed outwards, designed for the pick up of ambience and/or reverberation.
- Holophone, a barrier-type multichannel microphone with 5.1 outputs.

[1] Including my own book *Sound for Film and Television* from Focal Press.

- INA, a setup involving somewhat spaced directional mikes with the goal of 360° imaging in 5.1-channel surround.
- IRT cross, spaced cardioid mikes in a cross shape designed principally for the pick up of ambience and/or reverberation.
- Optimized Cardioid Triangle (OCT), a specific setup of a cardioid and two supercardioids in a spaced array, optimized for front imaging.
- Polyhymnia array consisting of 5 omnis at the angles of the loudspeaker channels.
- Sphere-type stereo microphone with added bidirectional pickups, using two M-S style matrixes to derive LF, RF, LR, RR; a center channel may be derived.
- Trinnov Array, an array of eight omni mikes and postprocessing to produce multiple directional microphone channels having greater directivity than available from standard microphones, adjusted specifically for the angles of the 5.1 standard.
- Combinations of these, such as Fukada array for front imaging and Hamasaki square for surround.
- Using real rooms as reverberation chambers or reverberation devices to produce surround channels usually used for pre-existing recordings.

Before describing stereo and multichannel surround microphone arrays, some features of microphones are worth describing, in particular due to the relationship between the features and surround sound usage:

- Pressure (omni) microphones have more extended range at the low end than directional ones; all-directional mikes (bidirectional, cardioid, etc.) have less very low-frequency range (below ca. 50 Hz) since the pressure difference at low frequency in the sound field is less. An example of how this affects one setup is for the OCT microphone array where an omni, low-pass filtered at 40 Hz (attenuating the frequencies above 40 Hz for this microphone), is added to the channels of the directional microphones to "fill in" for their very low-frequency inadequacy.
- Virtually all-directional mikes suffer from proximity effect, boosting the mid-bass (50–400 Hz) when the sound source is close to the microphone (although the rule that the extreme bass is attenuated in directional mikes still applies). The reason for this is the spherical expansion of sound and its interaction with this type of receiver, particularly when observed at close distance. It is avoided in only a few special designs that usually claim this as a feature, such as the AKG C4500 B-BC. Some manufacturers data sheets may be disingenuous on this point, showing flatter response to a lower frequency than in all likelihood is true.

- The noise floor of omni pressure mikes is lower than that of similar directional pressure-gradient directional mikes by an audible amount in some natural acoustic recording situations.
- The noise floor of most microphones is set by a tradeoff between diameter—larger being better for noise performance—and off-axis response—smaller being better for a more uniform polar pattern with frequency. Microphones having similar diaphragm diameters will usually exhibit the same noise floor. However, several techniques may be used to decrease the noise.[2] In certain instances this is particularly significant, such as in using backwards-facing cardioids a long ways from an orchestra in a hall to pick up hall sound where the average sound level is low.
- All other things being equal, a larger diaphragm mike will have a lower noise floor than a similar smaller diaphragm mike. However, its polar pattern will be more affected by its size, varying more across frequency. The low noise floor thus offered makes a large-diaphragm microphone like the Neumann TLM 103 with its 7 dBA SPL equivalent noise floor very useful for distant pickup in halls, particularly when used backwards facing, and when equalized for best off-axis leakage (probably rolling off highs). Table 3-1 below summarizes these last three points.
- Omni pattern mikes have a narrowing polar pattern at high frequencies. This "defect" of omnis is recognized and turned into a feature with certain array setups, such as the Decca tree, where the increasing directivity with frequency of omni microphones is put to good use to better separate the parts of an orchestra. In fact the original microphone usually used in the Decca Tree arrangement, the Neumann M50, contains a hard sphere of 40 mm diameter flush with the 12 mm capsule to exaggerate this effect with a smooth high-frequency shelf and increasing directivity at high frequencies. In some cases a hard spherical ball may be added to a conventional omni with the microphone inserted into the ball and made flush with the diaphragm entrance surface to increase the presence range level and directivity. Also see Schoeps PolarFlex below.

---

[2]Normally microphone diaphragm mass and the tension to which it is stretched form a resonant system, which is damped acoustically down to flat from a peaked response by locating a damping system near the back of the diaphragm. This damping mechanism is usually achieved by drilling holes in the back plate of an electrostatic type (called condenser or capacitor mike in common usage) mike and placing the back plate in close proximity to the diaphragm. If one were instead to not damp the acoustical resonance, but rather to use an electrical dip filter in the microphone to remove the resonance, the noise level is reduced in the frequency range of the resonance. If this can be arranged to be in the most audible frequency region of 2–3 kHz, a very real improvement in audible signal-to-noise ratio may be achieved.

Table 3-1 Comparative Noise of Microphones Using Various Technologies and Diaphragm Areas

Microphone model, description	*Noise floor, equivalent SPL* in dBA*
Schoeps CCM 2 L/U Omni, 20 mm diameter case	11
Schoeps CCM 8 L/U bidirectional, 20 mm diameter case	18
Sennheiser MKH 20 Omni, 25 mm diameter case	10
Sennheiser MKH 30 bidirectional, 25 mm diameter case, dual diaphragm	13
Neumann TLM 103 Cardioid, 27 mm diaphragm	7
*dB re 20 μN/m^2.	

- A seldom used type, the wide cardioid, with a polar pattern between that of an omni and a cardioid, may be designed to have a very uniform polar pattern with frequency, and thus may be put to good use in applications where spaced omnis would normally be employed and uniformity of off-axis response is desired to be good.
- *Sanken CU-41*: This single microphone has properties that make it especially suitable for use as the main stereo pair of recordings for the ORTF and X-Y methods. The microphone consists of two cardioid transducers, one large and one small, with a crossover between them, arranged closely together in one body. This arrangement makes it possible to keep the on- and off-axis response more uniform over the audible frequency range. Since ORTF and X-Y setups record the center of the stereo sound field off the axis of the main microphones, having available a microphone with especially good off-axis response is useful. Also, the two-way approach permits the specified bass response of this cardioid microphone to be unusually flat ±1 dB down to 20 Hz.
- *Schoeps PolarFlex*: This is a system comprised of two microphone pairs, each pair consisting of an omni and a bidirectional, or two cardioids; and a processing unit that allows the microphones to be combined into a stereo pair with various composite polar patterns. In particular not only does it offer adjustability of polar patterns electrically over the full range of first-order patterns (omni, bidirectional, cardioid, supercardioid), but it can also produce differing polar patterns in three different frequency ranges. So for example what is a cardioid at mid-frequencies can become an omni at low frequencies, and thus extend the frequency range, and a supercardioid at high frequencies, offering better isolation.

In the following, the various techniques are first reviewed briefly for stereo use, and then extending them to multichannel is covered.

Pan Pot Stereo

Panning multiple microphones into position to produce a constructed sound field is probably the most widely used technique for popular music, and complex events like sports or television specials. In pop music, this technique is associated with multitrack recording, and with the attendant capabilities of overdubbing and fine control during mixdown. In "event" sound, using many microphones with a close spacing to their sources means having more control over individual channels than the other methods offer. Although 100% isolation is unlikely in any practical situation this method still offers the most isolation (Fig. 3-1).

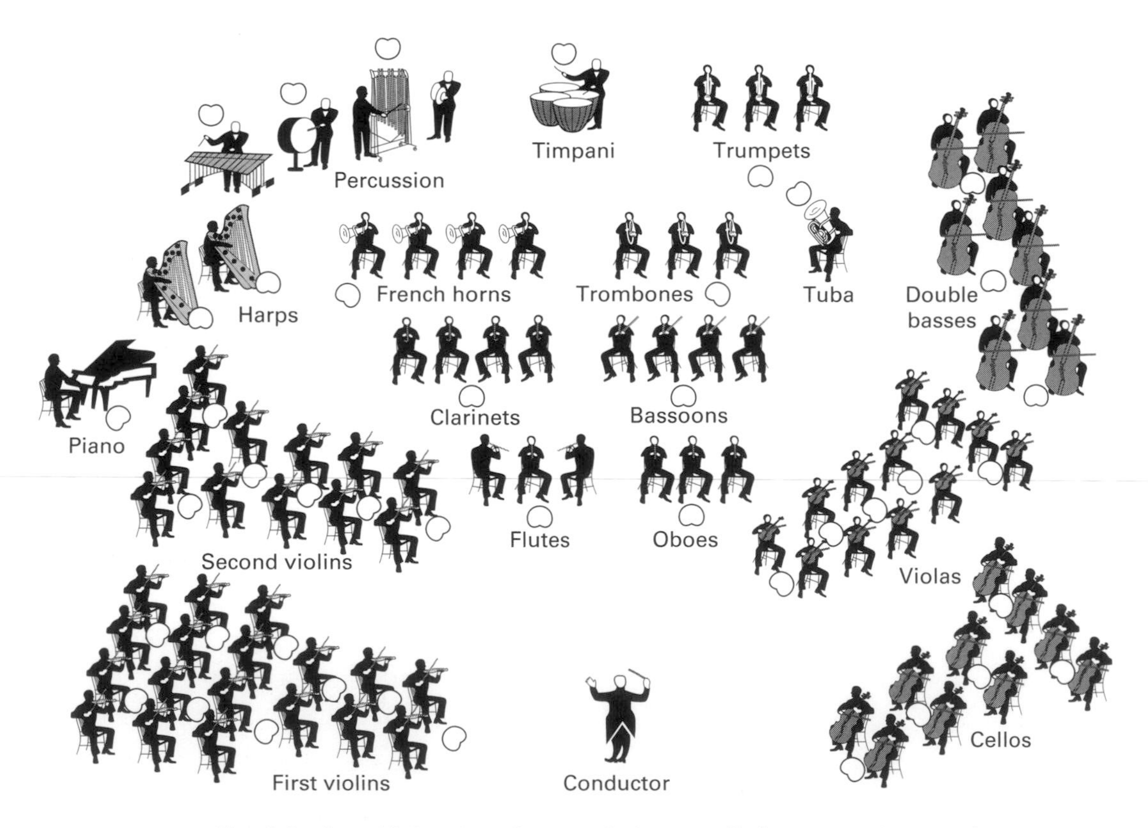

Fig. 3-1 A multiple microphone technique applied to a symphony orchestra involves miking individual or small numbers of instruments with each microphone to obtain maximum control over balance in mixing. Difficulties include capturing the correct timbre for each instrument, when the microphones are so close.

Some of the considerations in the use of multiple pan-potted microphones in either a stereo or multichannel context are:

- The relationship between the microphone and the source is important (as with all techniques, but made especially important in multimiking due to the close spacing used). Many musical instruments, and speech, radiate differing frequency ranges with different strengths in various directions. This is what makes microphone placement an art rather than a science, because a scientific approach would attempt to capture all of the information in the source. Since most sources have a complex spatial output, many microphones would be needed, say organized facing inward on a regular grid at the surface of a sphere with the source at the center. This method "captures" the 3-D complexity of sources, but it is highly impractical. Thus, we choose one microphone position that correctly represents the timbre of the source. Professionals come to know the best position relative to each instrument that captures a sound that best represents that instrument. In speech, that position is straight ahead or elevated in front of the talker; the direction below the mouth at 3 ft sounds less good than above, due to the radiation pattern of typical voices. If such a position below a frame line for instance must be employed, say in a classical music context for a video including a singer, then the engineer should feel free to equalize the sound to match a better position of the microphone. One way to do this is to put a microphone in the best location at 45° overhead and at say 3 ft temporarily in rehearsal, and one at the required position, and equalize the required position to match that of the good location by a/b comparison of the two positions by ear. This will usually mean taking out some chestiness by equalizing the range around, say, 630 Hz, with a broad dip of several dB.
- Often, microphones must be used close to instruments in order to provide isolation in mixing. With this placement, the timbre may be less than optimum, and equalization is then in order too. For instance, take a very flat microphone on axis such as a Schoeps MK2 omni. Place it several feet from a violin soloist, 45° overhead aimed at the source; this placement allows emphasis of this one violin in an orchestra. The sound is too screechy, with too much sound of rosin. The problem is not with the microphone, but rather with this close placement when our normal listening is at a distance—it really does sound that way at such a close position and at this angle. At a distance within a room, we hear primarily reflected sound and reverberation; the direct sound is well below the reflected and reverberant sound in level where we listen. What we actually hear is closer to an amalgamation of the sound of the violin at all angles, rather than the one that the close miking emphasizes.

While we need such a position to get adequate direct sound from the violin, suppressing the other violins, the position is wrecking the timbre. Thus, we need to equalize the microphone for the position, which may mean use of a high-frequency shelving equivalent down −4 dB at 10 kHz, or if the violin sounds overly "wirey," a broad dip of 2–3 dB centered around 3 kHz. The high-frequency shelf mimics the air losses that occur between nearby listening and listening at a substantial difference in a reverberant room. The presence range dip helps put the violin into the more distant perspective too.

- The unifying element in pan pot recordings is often reverberation. Although there are a number of specific multichannel reverberators on the market, whether in the form of separate hardware or software for digital audio workstations, a work-around if you don't have a multichannel device is the use of several stereo reverberators, with the returns of the various devices sent to the multiple channels, starting with left, right, left surround, and right surround as the most important. This is covered in Chapter 4 on Mixing.
- This method is criticized by purists for its lack of "real" stereo. However, note that the stereo they promote is coincident-miking, with its level difference only between the channels for the direct sound. (As a source moves across the stereo field with a coincident microphone technique it starts in one channel, then the other channel fades up, and then the first channel fades down, all because we are working around the polar pattern of the microphones. Sounds like pan potting to me!)

Spaced Omnis

Spaced microphone stereo is a technique that dates back to Bell Labs experiments of the 1930s. By recording a set of spaced microphones, and playing over a set of similarly spaced loudspeakers, a system for stereo is created wherein an expanding bubble of sound from an instrument is captured at points by the microphones, then supplied to the loudspeakers that continue the expanding bubble (Fig. 3-2). This "wavefront reconstruction" theory works by physically recreating the sound field, although the simplification from the desired infinite number of channels to a practical three results in some problems, discussed in the appendix on psychoacoustics. It is interesting that contemporary experiments, especially from researchers in Europe, continue along the same path in reconstructing sound fields. Considerations in the use of spaced microphones are:

- One common approach to spaced microphones is the "Decca tree." This setup uses three typically large-diaphragm omnidirectional

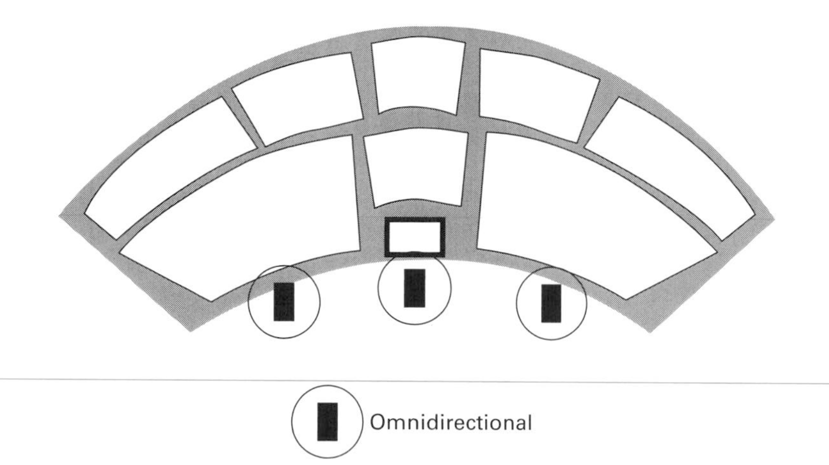

Fig. 3-2 Spaced omnis is one method of recording that easily adapts to 5.1-channel sound, since it is already commonplace to use three spaced microphones as the main pickup. With the addition of hall ambience microphones, a simple 5.1-channel recording is possible, although internal balance within the orchestra is difficult to control with this technique. Thus it is commonplace to supplement a basic spaced omni recording with spot microphones.

microphones arranged on a boom located somewhat above and behind the conductor of an orchestra, or in a similar position to other sound sources. The three microphones are spaced along a line, with the center microphone either in line, or slightly in front of (closer to the source), the left and right microphones. The end microphones are angled outwards.

- Spacing too close together results in little distinction among the microphone channels since they are omnidirectional and thus only small level and timing differences would occur. Spacing the microphones too far apart results in potential audible timing differences among the channels, up to creating echoes. The microphone spacing is usually adjusted for the size of the source, so that sounds originating from the ends of the source are picked up nearly as well as those from the center. An upper limit is created on source size when spacing the microphones so far apart would cause echoes. Typical spacing is in the range of 10–30 ft across the span from left to right, but I have used a spacing as large as 60 ft in a five front channel setup when covering a football game half-time band activities without echo problems.
- Spaced microphones are usually omnis, and this technique is one that can make use of omnis (coincident techniques require directional microphones, and pan-potted stereo usually uses directional mikes for better isolation). Omnidirectional microphones are pressure-responding microphones with frequency response that

typically extends to the lowest audible frequencies, whereas virtually all pressure-gradient microphones (all directional mikes have a pressure-gradient component) roll off the lowest frequencies. Thus, spaced omni recordings exhibit the deepest bass response. This can be a blessing or a curse depending on the nature of the desired program material, and the noise in the recording space.

- Spaced microphones are often heard, in double blind listening against coincident and near-coincident types of setups, to offer a greater sense of spaciousness than the other types. Some proponents of coincident recording ascribe this to "phasiness" rather than true spaciousness, but many people nonetheless respond well to spaced microphone recordings. On the other hand, good imaging of source locations is not generally as good as with coincident or near-coincident types of miking.
- Spaced microphone recordings produce problems in mixdown situations, including those from 5.1 to 2 channels, and 2 channels to mono. The problem is caused by the time difference between the microphones for all sources except those exactly equidistant from all the mikes. When the microphone outputs are summed together in a mixdown, the time delay causes multiple frequencies to be accentuated, and others attenuated. Called a "comb filter response," the frequency response looks like a comb viewed sideways. The resulting sound is a little like Darth Vader's voice, because the processing that is done to make James Earl Jones sound mechanical is to add the same sound repeated 10 ms later to the original; this is a similar situation to a source being located at an 11 ft difference between two microphones.

Coincident and Near-Coincident Techniques

Crossed Figure-8

Coincident and near-coincident techniques originated with the beginnings of stereo in England also in the 1930s. The first technique named after its inventor Blumlein used crossed Figure-8 pattern bidirectional microphones. With one Figure-8 pointed 45° left of center, and the other pointed 45° right of center, and the microphone pickups located very close to one another, sources from various locations around the combined microphones are recorded, not with timing differences because those have been essentially eliminated by the close spacing, but with simply level differences. A source located on the axis of the left-facing Figure-8 is recorded at full level by it, but with practically no direct sound pickup in the right-facing Figure-8, because its null is pointed along the axis of the left-facing mike's highest output.

For a source located on a central axis in between left and right, each microphone picks up the sound at a level that is a few dB down from pickup along its axis. Thus, in a very real way, the crossed Figure-8 technique produces an output that is very much like pan-potted stereo, because pan pots too produce just a variable level difference between the two channels.

Some considerations of using crossed Figure-8 microphones are:

- The system makes no level distinction between front and back of the microphone set, and thus it may have to be placed closer than other coincident types, and it may expose the recording to defects in the recording space acoustics.
- The system aims the microphones to the left and right of center; for practical microphones, the frequency response at 45° off the axis might not be as flat as that on axis, so centered sound may not be as well recorded as sound on the axis of each of the microphones.
- Mixdown to mono is very good since there is no timing difference between the channels—a strength of the coincident microphone methods.

This system is probably not as popular as some of the other coincident techniques due to the first two considerations above (Fig. 3-3).

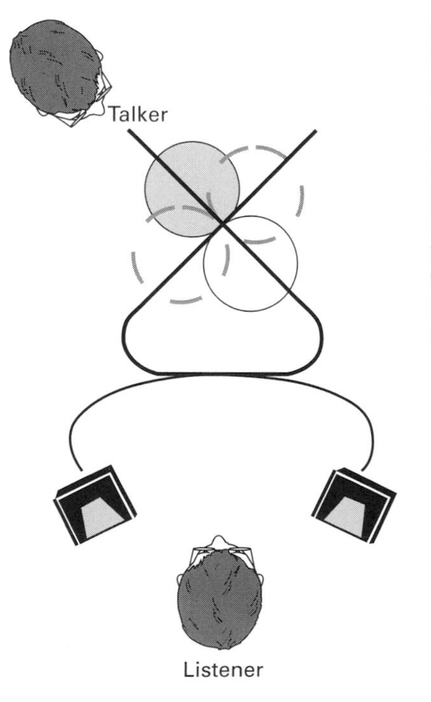

Fig. 3-3 As the talker speaks into the left microphone, he is in the null of the right microphone, and the left loudspeaker reproduces him at full level, while the right loudspeaker reproduces him at greatly reduced level. Moving to center, both microphones pick him up, and both loudspeakers reproduce his voice. The fact that the microphones are physically close together makes them "coincident" and makes the time difference between the two channels negligible.

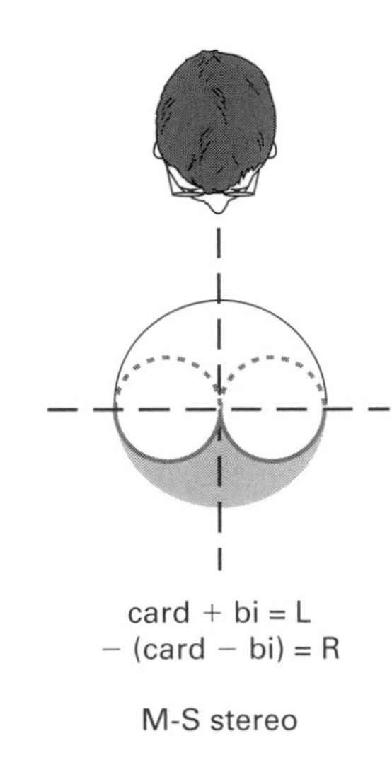

Fig. 3-4 M-S Stereo uses a forward-firing cardioid, hypercardioid, or even shotgun, and a side firing bidirectional microphone. The microphone outputs are summed to produce a left channel, and the difference is taken and then phase flipped to produce a right channel. The technique has found favor in sound effects recording.

M-S Stereo

The second type of coincident technique is called M-S, for mid-side (Fig. 3-4). In this, a cardioid or other forward-biased directional microphone points toward the source, and a Figure-8 pattern points sideways; of course, the microphones are co-located. By using a sum and difference matrix, left and right channels can be derived. This works because the front and back halves of a Figure-8 pattern microphone differ from each other principally in polarity: positive pressure on one side makes positive voltage, while on the other side, it makes a negative voltage. M-S exploits this difference. Summing a sideways-facing Figure-8 and a forward-firing cardioid in phase produces a left-forward facing lobe. Subtracting the two produces a right-forward facing lobe that is out of phase with the left channel. A simple phase inversion then puts the right channel in phase with the left. M-S stereo has some advantages over crossed Figure-8 stereo:

- The center of the stereo field is directly on the axis of a microphone pickup.
- M-S stereo distinguishes front and back; back is rejected because it is nulled in both the forward-facing cardioid, and in the sideways facing Figure-8, and thus a more distant spacing from orchestral sources can be used than that of Blumlein, and/or less emphasis is placed on the acoustics of the hall.
- Mixdown to mono is just as good as crossed Figure-8 patterns, or perhaps even better due to the center of the stereo field being on axis of the cardioid.

- M-S stereo is compatible with the Dolby Stereo matrix; thus, it is often used for sound effects recordings.

X-Y Stereo

X-Y is a third coincident microphone technique that uses crossed cardioid pattern microphones, producing left and right channels directly. It shares some characteristics of both the crossed Figure-8 and the M-S techniques. For instance, it:

- Distinguishes front from back by the use of the cardioid nulls.
- Has the center of the stereo sound field off the axis of either microphone; due to this factor that it shares with the crossed Figure-8 microphones, it may not be as desirable as M-S Stereo.
- X-Y stereo requires no matrix device, so if one is not at hand it is a quick means to coincident stereo, and most studios have cardioids at hand to practice this technique.
- Summing to mono is generally good.

All of the standard coincident techniques suffer from a problem when extended to multichannel. Microphones commonly available have first-order polar patterns (omni, bidirectional, and all variations in between: wide cardioid, cardioid, hypercardioid, supercardioid). Used in tight spacing, these exhibit not enough directivity to separate L/C/R sufficiently. Thus many of the techniques to be described use combinations of microphone spacing and barriers between microphone pickups to produce adequate separation across the front sound stage.

Near-Coincident Technique

Near-coincident techniques use microphones at spacings that are usually related to the distance between the ears. Some of the techniques employ obstructions between the microphones to simulate some of the effects of the head. These include:

- *ORTF stereo*: A pair of cardioids set at an angle of 110° and a spacing equal to ear spacing. This method has won over M-S, X-Y, and spaced omnis in stereo blind comparison tests. At low frequencies where the wavelength of sound in air is long, the time difference between the two microphones is small enough to be negligible, so the discrimination between the channels is caused by the different levels due to the outwardly aimed cardioid polar patterns. At high frequencies the difference is caused by a combination of timing and level, so the result is more phasey than completely coincident microphones, but not so much as to cause too much trouble (Fig. 3-5).

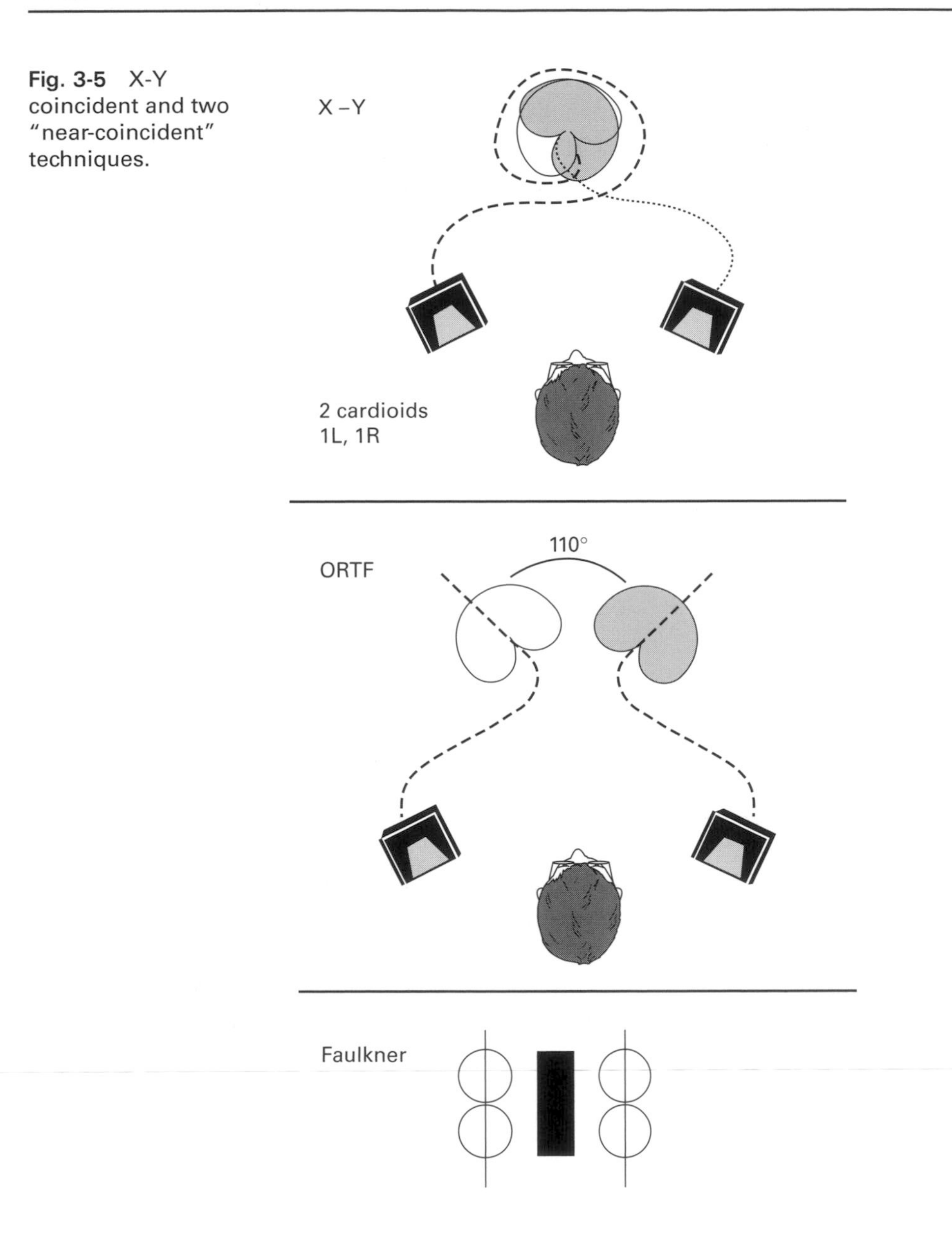

Fig. 3-5 X-Y coincident and two "near-coincident" techniques.

- *Faulkner stereo*: UK recording engineer Tony Faulkner uses a method of two spaced Figure-8 microphones with their spacing set to ear-to-ear distance and with their pickup angle set to straight forward and backward. A barrier is placed between the microphones.
- *Sphere microphone*: Omni microphone capsules are placed in a sphere of head diameter at angles where ears would be. The theory is that by incorporating some aspects of the head (the sphere),

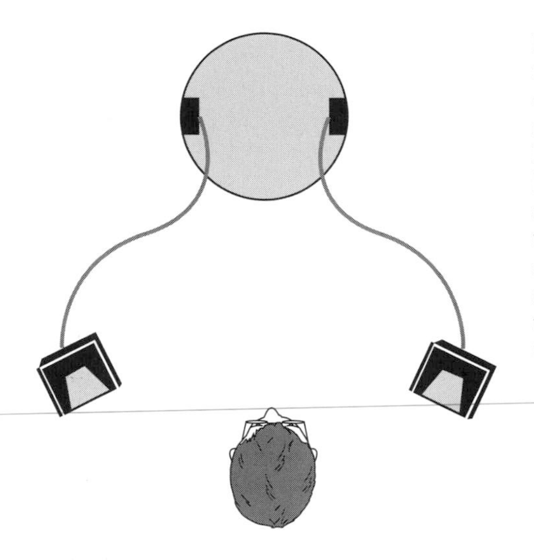

Fig. 3-6 A sphere microphone mimics some of the features of dummy head binaural, while remaining more compatible with loudspeaker playback. Its principles have been incorporated into a 5.1-channel microphone, by combining the basic sphere microphone with an M-S system for each side of the sphere, and deriving center by a technique first described by Michael Gerzon.

while neglecting the pinna cues that make dummy head recordings incompatible with loudspeaker listening, natural stereo recordings can be achieved. The microphone is made by Schoeps (Fig. 3-6).

Near-coincident techniques combine some of the features of both coincident and spaced microphones. Downmixing is likely to be better than with more distantly spaced mikes, while spaciousness may be better than that of strictly coincident mikes. A few comparison studies have been made. In these studies, multiple microphone techniques are recorded to a multitrack tape machine, then compared in a level-matched, blind listening test before experts. In these cases, the near-coincident technique ORTF has often been the top vote-getter, although it must be said that any of the techniques have been used by record companies and broadcast organizations through the years to make superb recordings.

Binaural

The final stereo microphone to be considered is not really a stereo microphone at all, but rather a binaural microphone called a dummy head. Stereo is distinguished from binaural by stereo being aimed at loudspeaker reproduction, and binaural at headphone reproduction. Binaural recording involves a model of the human head, with outer ears (pinna), and microphones placed either at the outer tip of the simulated ear canal, or terminating the inside end of an artificial ear canal. With signals from the microphones supplied over headphones, a more or less complete model of the external parts of the human hearing system is produced. Binaural recordings have the following considerations:

- This is the best system at reproducing a distance sensation from far away to very close up.
- Correctly located sound images outside the head are often achieved for sound directly to the left, right, and overhead.
- Sound images intended to be in front, and recorded in that location, often can sound "inside the head;" this is thought to be due to the non-individualized recording (through a standard head and pinnae, not your own) that binaural recording involves, and the fact that the recording head is fixed in space, while we use small head movements to "disambiguate" external sound locations in a real situation.
- Front/back confusion is often found. While these can occur in real acoustic spaces, they are rare and most people have probably never noticed them. Dynamic cues of moving one's head a small amount disambiguate front from back usually in the real world, but not with dummy head recordings.
- Binaural recordings are generally not compatible with loudspeaker reproduction, which is colored by the frequency response variations caused by the presence of the head twice, once in the recording, and once in the reproduction.
- Use of a dummy head for recording the surround component of 5.1-channel mixes for reproduction over loudspeakers has been reported—the technique may work as the left and right surround loudspeakers form, in effect, giant headphones.

Spot Miking

All of the techniques above, with the lone exception of pan-potted stereo, produce stereo from basically one point of view. For spaced omnis, that point of view may be a line, in an orchestral recording over the head and behind the conductor. The problem with having just one point of view is the impracticality of getting the correct perspective and timbre of all of the instruments simultaneously. That is why all of the stereo techniques may be supplemented with taking a page from close-miked stereo and use what are called spot or accent mikes. These microphones emphasize one instrument or group of instruments over others, and allow more flexible balances in mixing. Thus, an orchestra may be covered with a basic three-mike spaced omni setup, supplemented by spot mikes on soloists, woodwinds, timpani, and so forth. The level of these spot microphones will probably be lower in the mix than the main mikes, but a certain edge will be added of clarity for those instruments. Also, equalization may be added for desired effect. For instance, in the main microphone pickup of an orchestra, it is easy for tympani to sound too boomy. A spot mike on the tymp, with its bass rolled off, provides just the right "thwack" on attacks that sounds closer to what we actually hear in the hall.

One major problem with spot miking has been, until recently, that the spot mikes are located closer to their sources than the main mikes, and therefore they are earlier in time in the mix than the main ones. This can lead to their being hard to mix in, as adding them into the mix not only changes both the relative levels of the microphones, but also the time of arrival of the accented instrument, which can make it seem to come on very quickly—the level of the spot mikes becomes overly critical. This is one primary reason to use a digital console: each of the channels can be adjusted on most digital consoles in time as well as level. The accent mike can be set back in time to behind the main mikes, and therefore the precedence effect (see Chapter 6) is overcome.

On the other hand, Florian Camerer of Austrian ORF broadcasting reports that for main microphone setups that have narrow and/or vague frontal imaging, non-delayed spot mikes can be useful to set the direction through panning them into place and positive use of the precedence effect. Of course spot mike channels would use reverberation to blend them in. Such an example is a Decca Tree with 71 in. (180 cm) spacing between the left and right microphones, and the center microphone placed 43 in. (110 cm.) in front of the line formed by left on the right. The two recording angles of this array are 20° left to center, and 20° right to center. While spacious sounding, this is basically triple mono, with little imaging, and it is here where non-delayed spot mikes can be effectively employed.

Multichannel Perspective

There are two basically different points of view on perspective in multichannel stereo. The first of these seeks to reproduce an experience as one might have it in a natural space. Called the "Direct/Ambient" approach, the front channels are used mostly to reproduce the original sound sources, and the surround channels are used mostly to reproduce the sense of spaciousness of a venue through enveloping the listener in surround sound reproducing principally reverberation. Physical spaces produce reflections at a number of angles, and reverberation as mostly a diffuse field, from many angles, and surround sound, especially the surround loudspeakers, can be used to reproduce this component of real sound that is unavailable in 2-channel stereo.

The pros of the direct/ambient approach are that it is more like listening in a real space, and people are thus more familiar with it in everyday listening. It has one preferred direction for the listener to face, so it is suitable for accompanying a picture. The cons include the fact that when working well it is often not very noticeable to the man on the street. To demonstrate then the use of surround, the best way is to set up the system with equal level in all the channels (not being tempted to "cheat" the surround level higher to make it more noticeable), and then to find appropriate program material with a good direct-to-reverberant balance. Shutting off the surrounds abruptly shows what is lost, and is an effective demo. Most people react to this by saying, "Oh, I didn't know it was doing so much," and thus become educated about surround.

The second perspective is to provide the listener with a new experience that cannot typically be achieved by patrons at an event, an "inside the band" view of the world. In this view, all loudspeaker channels may be sources of direct sound. Sources are emitted all round one, and one can feel more immersed. Pros include a potentially higher degree of involvement than with the direct/ambient approach; but cons are that many people are frightened or annoyed at direct sound sources coming from behind them—they feel insecure. A Gallup poll conducted by the Consumer Electronics Association asked the requisite number of persons in a telephone poll needed to get reliable results on the population as a whole. About 2/3 of the respondents preferred what we've called the direct/ambient approach, while 1/3 preferred a more immersive experience. That is not to say that this approach should not be taken, as it is an extension of the sound art to have available this dimension of sound for artistic expression, but practitioners should know that the widespread audience may not yet be prepared for a fully surrounding experience.

Use of the Standard Techniques in Multichannel

Most of the standard techniques described above can be used for at least part of a multichannel recording. Here is how the various methods are modified for use with the 5.1-channel system.

Pan-potted stereo changes little from stereo to multichannel. The pan pot itself does get more complicated, as described in Chapter 4. The basic idea of close miking for isolation remains, along with the idea that reverberation provides the glue that lets the artificiality of this technique hang together. Pan-potted stereo can be used for either a Direct/Ambient approach, or an "in the band" approach. Some considerations in pan-potted multichannel are:

- Imaging the source location is perfect at the loudspeaker positions. That is, sending a microphone signal to only one channel permits everyone in the listening space to hear the event picked up by that microphone at that channel. Imaging in between channels is more fragile because it relies on phantom images formed between pairs of channels. One of the difficulties is that phantom images are affected greatly by the precedence effect, so the phantom images move with listening location. However, increasing the number of channels decreases the angular difference between each pair of channels and that has the effect of widening the listening "sweet spot" area.

- The quality of phantom sound images is different depending where the source is on the originating circle. For 5.1-channel sound, across left, center, and right, and, to a lesser extent, again across the back between left surround and right surround, phantom images are formed in between pairs of loudspeakers such that imaging is relatively good in these areas. Panning part way between left and left surround, or right and right surround, on the other hand, produces very poor results, because the frequency response in your ear canal from even perfectly matched speakers is quite different for L and LS channels, due to Head Related Transfer Functions (HRTFs). See Chapter 6 for a description of this effect. The result of this is that panning halfway between L and LS electrically results in a sound image that is quite far forward of halfway between the channels, and "spectral splitting" can be heard, where some frequency components are emphasized from the front channel, and others from the surround channel. The sound "object" splits in two, so a pan from a front to a surround speaker location starts by hearing new components in the frequency range fade in on the surround channel, then fade out on the front channel; the sound image "snaps" at some point during the pan to the surround speaker location. By the way, one of the principal improvements in going from 5.1 to 10.2-channel sound is that the Wide channels, ±60° from front, help to "bridge the gap" between left at 30° and left surround at 100–120° and pans from left through left wide to left surround are greatly improved so that imaging all round becomes practical, and vice versa on the right.
- Reverberation devices need multichannel returns so that the reverberation is spatialized. Multichannel reverberators will supply multiple outputs. If you lack one, a way around this is to use two stereo reverberation devices fed from the same source, and set them for slightly different results so that multiple, uncorrelated outputs are created for the multiple returns. The most effective placement for reverberation returns is left, right, left surround, and right surround, neglecting center, for psychoacoustic reasons.
- Pan-potted stereo is the only technique that supports multitrack overdubbing, since the other techniques generally rely on having the source instruments in the same space at the same time. That

is not to say that various multichannel recordings cannot be combined, because they can; this is described below.

Most conventional stereo coincident techniques are not directly useful for multichannel without modification, since they are generally aimed at producing just two channels. A major problem for the coincident techniques is that the microphones available on the market, setting aside one specialized type for a moment, are "first-order" directionality polar-pattern types (bidirectional, wide cardioid, hypercardioid, supercardioid). First-order microphones, no matter how good, or of which directionality, are simply too wide to get adequate isolation among left, center, and right channels when used in coincidence sets, so either some form of spacing must be used, or more specialized types employed. Several partial solutions to this problem are described below.

There are some specialized uses, and uses of coincident techniques as a part of a whole, that have been developed for multichannel. These are:

- Use of an ORTF near-coincident pair as part of a system that includes outrigger mikes and spot mikes. The left and right ORTF pair microphones are panned just slightly to the left and slightly to the right of the center channel in mixing. See a description at the end of this chapter developed by John Eargle of how this system can work to make stereo and multichannel recordings simultaneously.
- Combining two techniques, the sphere mike and M-S stereo, results in an interesting 4-channel microphone. This system developed by Jerry Bruck uses a matrix to combine the left omni on a sphere with a left Figure-8 mike located very close to the omni and facing forward and backward, and vice versa for the right. This is further described under Special Microphones for 5.1-channel recordings.
- Extending the M-S idea to 3-D is a microphone called the Sound Field mike; it too is described below.

Binaural dummy head recording is also not directly useful for multichannel work, but it can form a part of an overall solution, as follows:

- Some engineers report using a dummy head, placed in the far field away from instruments in acoustically good studios, and sending the output signals from L and R ears to LS and RS channels. Usually, dummy head recordings, when played over loudspeakers, show too much frequency response deviation due to the HRTFs involved. In this case, it seems to be working better than in the past, possibly because supplying the signals at such angles to the head results in binaural imaging working, as the LS and RS channels operate as giant headphones.

- Binaural has been combined with multichannel and used with 3-D IMAX. 3-D visual systems require a means to separate signals to the two eyes. One way of doing this is to use synchronized "shutters" consisting of LCD elements in front of each eye. A partial mask is placed over the head, holding in place the transmissive LCD elements in front of each eye. Infrared transmission gives synchronizing signals, opening one LCD at a time in sync with the projector's view for the appropriate eye. In the mask are headphone elements located close by the ears, but leaving them open to external sound, too. The infrared transmission provides two channels for the headphone elements. In the program that I saw, the sounds of New York harbor including seagulls represented flying overhead in a very convincing way. Since binaural is the only system that provides such good distance cues, from far distant to whispering in your ear, there may well be a future here, at least for specialty venues. I thought the IMAX presentation was less successful when it presented the same sound from the headphone elements as from the center front loudspeaker: here timing considerations over the size of a large theater prevent perfect sound sync between external and binaural fields, and comb filters resulted at the ear. The pure binaural sound though, overlaid on top of the multichannel sound, was quite good.

Surround Technique

Perhaps the biggest distinguishing feature of multichannel in application to stereo microphone technique is the addition of surround channels. The reason for this is that in some of the systems, a center microphone channel is already present, such as with spaced omnis, and it is no stretch to provide a separate channel and loudspeaker to a microphone already in use. In other methods, it is simple to derive a center channel. Surround channels, however, have got to have a signal derived from microphone positions that may not have been used in the past.

Surround Microphone Technique for the Direct/Ambient Approach

Using the direct/ambient approach, pan pot stereo, spaced omnis, and coincident techniques can all be used in the frontal stage LCR stereo. However, coincident microphone techniques should be done with knowledge of the relative pickup angles of the microphones in use; achieving good enough isolation across LCR is problematic with normal microphones. In playback, the addition of the center channel solidifies the center of the stereo image, providing greater freedom in listening position than stereo, and a frequency response that does not suffer from crosstalk-induced dips in the 2 kHz region and above described in Chapter 6. The surround loudspeaker channels, on the other hand, generally require more microphones. Several approaches to surround channel microphones, often just one pair of spaced microphones, for the direct/ambient approach are:

- In a natural acoustic space like a concert hall, omnis can be located far enough from the source that they pick up mostly the reverberation of the hall. According to Jonathan Stokes[3] it is difficult to give a rule of thumb for the placement of such microphones, because the acoustics of real halls varies considerably, and many chosen locations may show up acoustic defects in the hall. That having been said, it is useful to give as a starting point something on the order of 30–50 ft from the main microphones. Locating mikes so far from the source could lead to hearing echoes, as the direct sound leakage into the hall mikes is clearly delayed compared to the front microphones. In such a case it is common to use audio delay of the main microphones to place them closer in time to the hall microphones, or to adjust the timing in postproduction on a digital audio workstation. This alone makes a case for having a digital console with time delay on each channel, so long as enough is available, to prevent echoes from distantly spaced microphones.

Of course, in live situations if the time delay is large enough to accommodate distantly spaced hall microphones, the delay could be so large that "lip sync" would suffer in audio for film or video applications. Also, performers handle time delay to monitor feeds very poorly, so stage monitors must not be delayed.

- An alternate to distantly spaced omnis is to use cardioids, pointed away from the source, with their null facing the source, to deliver a higher ratio of reverberation to direct sound. Since this is so they can be used closer to the source, perhaps at 1/2 the distance, of an equivalent omnidirectional microphone. Such cardioids will probably receive a lower signal level than any other microphone discussed, so the microphone's self noise, and preamplifier noise, become important issues for natural sound in real spaces using this approach. Nevertheless, it is a valid approach that increases the hall sound and decreases the direct sound, something often desirable in the surround channels. One of the lowest noise cardioids is the Neumann TLM 103.

[3]A multiple Grammy award winning classical music engineer with experience in many concert halls.

- The IRT cross, an arrangement of four spaced cardioids facing outwards arranged in a square about 10 in. (25 cm) on a side at 45° incidence to the direct sound has been found to be a useful arrangement for picking up reverberation in concert halls; one could see it almost as double ORTF. Also, it may be used for ambiences of sound effects and other spatial sound where imaging is not the first consideration, but spaciousness is. The outputs of the four microphones are directed to left, right, left surround, and right surround loudspeakers. A limitation is that some direct sound will reach especially the front facing microphones and pollute its use as a pickup of principally reverberation.
- The Hamasaki square array is another setup useful in particular for hall ambience. In it, four bidirectional mikes are placed in a square of 6–10 ft on a side, with their nulls facing the main sound source so that the direct sound is minimized, and their positive polarity lobes facing outwards. The array is located far away and high up in the hall to minimize direct sound. The front two are routed to L and R and the back two are routed to LS and RS. Side-wall reflections and the side component of reverberation are picked up well, while back wall echoes are minimized. In one informal blind listening test this array proved the most useful for surround.

Surround Microphone Technique for the Direct Sound Approach

For perspectives that include "inside the band" the microphone technique for the surround channels differs little from that of the microphones panned to the front loudspeakers. Mixing technique optimally may demonstrate some differences though, due to the different frequency response in the ear canal of the surround speakers compared to the fronts. Further information about this is in Chapters 4 and 6.

Special Microphones Arrays for 5.1-Channel Recordings

A few 5.1-channel specific microphone systems have appeared on the market, mostly using a combination of the principles of the various stereo microphone systems described above extended to surround. There are also a few models that have special utility in parts of a 5-channel recording, and setups using conventional microphones but arranged specifically for 5.1-channel recording. They are, in alphabetical order:

- *Double M-S*: Here three microphone capsules located close to one another, along with electronic processing, can produce left/center/

right/left surround/right surround outputs from one compact array. Two cardioids or preferably super- or hyper-cardioids face back to back, with one aimed at the primary sound source and the other away from it. A bidirectional mic. is aimed perpendicular to these first two, with its null in their plane of greatest output. By sum and difference matrixing as described above, the various signals can be derived. What keeps this from being an ideal technique is the fact that the angle of acceptance of each of the capsules overlaps one another too much due to the use of first-order microphones. However as a single-point recording system with very good compatibility across mono, stereo, and surround, this technique is unsurpassed.

- *Fukada array*: This setup for front channel microphones developed at NHK is derived from the Decca Tree, but the omnis are replaced with cardioids with specific distances and angles. The reason to change from omni to cardioid is principally to provide better isolation of front channel direct sound from ambient reverberation. Two cardioids are placed at the ends of a 6-ft long line, facing outwards. Thus these cardioids must have good off-axis performance since their 90° axis is aimed at the typical source. This polar pattern requirement dictates that they should be small-diaphragm microphones. A third caridioid is placed on a perpendicular line bisecting the first 5 ft from the first line and facing the source. This setup won rather decisively an informal blind listening test comparison among various of the types.[4] However, the authors are clear to call their paper informal in its title, and the number of sources is limited and various microphone models were employed for the various setups thus not quite making a comparison among apples but in some ways apples to oranges. Nevertheless this result is intriguing, as several authors have written against spaced microphones arrays of any sort. The Fukada array would normally be supplemented by backwards-facing cardioids spaced away from the array, an IRT cross, or a Hamasaki square, for surround. Of these, the Hamasaki square produced the best results in the informal listening test.
- *Holophone Global Sound Microphone system (multiple models)*: This consists of a set of pressure microphones flush mounted in a dual-radius ellipsoid, and a separate pressure microphone interior to the device for the Low Frequency Enhancement (LFE) channel. Both wired and wireless models have been demonstrated. Its name should not be confused with Holophonics, an earlier dummy head recording system.

[4]Rafael Kassier, Hyun-Kook Lee, Tim Brookes, and Francis Rumsey, "An Informal Comparison among Surround Sound Microphone Techniques," AES Preprint 6429.

- *Ideal Cardioid Arrangement INA-3 and INA-5*: Three (for front imaging) or five (for front imaging plus surround) typically cardioids are placed on arms and spaced in directions like those of the ITU Rec. 775 loudspeaker array. One example is the Gefell INA-5. Another is the Brauner ASM-5 microphone system and Atmos 5.1 Model 2600 matching console. The Brauner/Atmos system consists of an array of five microphones that additionally offers electrically adjustability from omnidirectional through Figure-8. Their mechanical configuration is also adjustable within set limits, and the manufacturer offers preferred setup information. The console provides some special features for 5.1-channel work, including microphone pattern control, LFE channel extraction, and front all-pass stereo spreading.
- *OCT array*: This array was developed by Günther Theile at the IRT. This array consists of two hypercardioids mounted at the ends of a line and facing outwards, and a cardioid capsule mounted on a line centered on and perpendicular to the first. The dimensions can be varied to meet a requirement that may be set for included angle. The objective is to produce better phantom images at left-center and right-center locations than other array types, since most setups overlap the outputs of the various microphones employed in the array so much as to make the half-images "mushy." Very good hypercardioids must be used since the principal sound is at 90° off axis and this requirement dictates the use of small-diaphragm microphones. In addition a pressure mike may be added in the mix for below 40 Hz response by low-pass filtering an omni included with the array, making a total of four mikes and mike channels to record this front array. It would normally be used with spaced backwards-facing cardioids, or an IRT cross or Hamanaki square array located behind it relative to the source as the channels to employ for surround envelopment and spaciousness. Thus a total of eight recorded channels would be used. The OCT array did not do very well in the informal listening test referenced above, however, it was realized in that case with large-diaphragm microphones that are not good in this service, so can't be counted too much against this rig. I have heard this setup in one AES conference demonstration and it seemed to image sources moving across the front sound stage quite well. In fact, its reason for being is this sharp frontal imaging, of particular use in small-ensemble music recording, and in some kinds of stereo sound effects recordings that must match to picture.
- *Sanken WMS-5*: The 5-channel microphone, which is a variation on double M-S. A short shotgun mike is used for center front, and two cardioids and a bidirectional mike are combined with it and an M-S style matrix to produce a 5-channel output. Note that the shotgun devolves to a hypercardioid at medium and low frequencies,

Fig. 3-7 The Schoeps KFM-360 Surround Microphone.

so the system is first-order up to the frequency where the shotgun gets more directional than a hypercardioid.

- *Schoeps KFM-360 Surround Microphone and DSP 4 electronics*: This system consists of a stereo sphere microphone (one of the barrier techniques), supplemented with two pressure-gradient microphones placed in near proximity to the two pressure microphones of the sphere (see Fig. 3-7). An external sum and difference matrix produces 5 channels in a method related to M-S stereo. The center channel is derived from left and right using a technique first described by Michael Gerzon.
- *SoundField microphones, several models*: SoundField microphones consist of a tightly spaced array of directional transducers arranged in a tetrahedron. Electronic processing of their outputs produces four output signals corresponding to Figure-8 pattern microphones pointed left–right, front–back, up–down, and an omnidirectional pressure microphone. The microphone is said to capture all the aspects of a sound field at a point in space; however, only first-order separation is available among the channels. The "B format" 4-channel signal may be recorded for subsequent postprocessing after recording, including steering. A 5-channel derivation is available from a model of a processing box for B format to 5.1 channels.

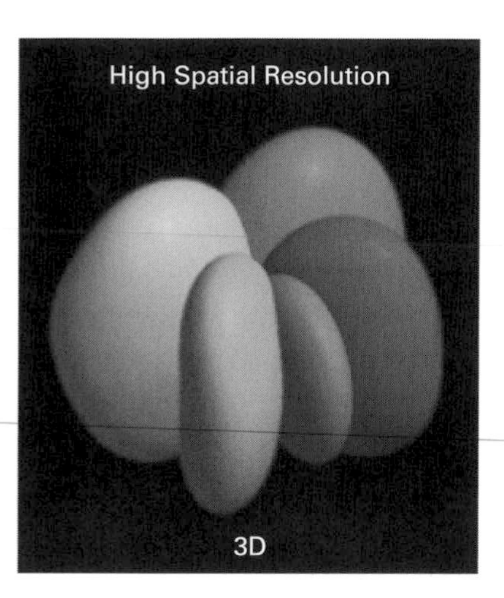

Fig. 3-8 A 3-D diagram of the pickup pattern of one Trinnov array. Center is the small shape down and to the right; left and right straddle it; the surrounds are the two larger lobes to the rear.

- *Trinnov array*: A set of omnidirectional microphones used with DSP postprocessing and high-order mathematical functions can produce the effect of high-order lobes, pointed in various directions and with adjustable overlaps, and thus solving a fundamental problem. An 8-microphone array is designed specifically for the angles of the 5.1-channel system. This is an attempt to make an effectively coincident recording but with higher-order functions to permit greater separation of the channels (Fig. 3-8).

It must be said that the field of microphone techniques for surround sound, although not growing perhaps as quickly now as in the period from 1985 to 2000, is still expanding. Among the authors prominent in the field, Michael Williams has produced studies of the recording included angle for various microphone setups and so forth. His writings along with that of others are to be found on the www.aes.org/e-lib web site. A search of the term "surround microphone" at this site revealed 58 examples as this is being written (Fig. 3-9).

In the future, higher-order gradient microphones may become available that would make single-point multichannel pickup improvements through sharpening the polar pattern of the underlying component microphones. However, the problem is not only commercial for such an array, but theoretical in that signal-to-noise ratio problems of higher-order microphones have not yet been addressed.

Combinations of Methods

In some kinds of program-making, the various techniques described above can be combined, each one used layered over others. For instance, in recording for motion pictures, dialogue is routinely recorded

Fig. 3-9 Florian Camerer of Austrian Television shows his devotion to location surround sound recording.

monaurally, then used principally in the center channel or occasionally panned into other positions. Returns of reverberation for dialogue may include just the screen channels, or both the screen and surround channels, depending on the desired point of view of the listener. In this way, dialogue recording is like standard pan pot recording. Many kinds of specific sound effect recordings employ the same technique, such as Foley recording (watching the picture and matching sound effects to it), and "hard effects," sound effects of specific items that you generally see.

Other sound effect recordings where spaciousness is more important are made with an M-S technique. One reason for this is the simplicity in handling. For instance, Schoeps has single hand-held shock mounts and windscreens for the M-S combination that makes it very easy to use. Another reason is the utility of M-S stereo in a matrix surround (Dolby Stereo) system, since in some sense; the M-S system matches the amplitude-phase matrix system. A third reason is that the 2-channel outputs of the M-S process can be recorded on a portable 2-channel recorder, rather than needing a portable multichannel recorder. When used in discrete systems like 5.1, it may be useful to decode the M-S stereo into LCRS directions using a surround matrix decoder, such as Dolby's SDU-4 (analog) or 564 (digital).

Ambience recordings are usually 2-channel stereo spatialized into multiple channels either by the method described for M-S sound effects above, or by other methods described in Chapter 4.

Orchestral music recordings often use the Decca tree spaced omni approach, with accent microphones for solos. The original multitrack recording will be pre-mixed down and sent to postproduction mixing as an L/C/R/LS/RS/LFE master. Note that the order of this list does not necessarily match the order of tracks on the source machine. AES TD-1001 specifies an 8-track master be laid out L, R, C, LFE, LS, RS, L (stereo), R (stereo), although many other track layouts are in use. Studios must consult one another for interchange in this area.

So a motion picture sound track may employ a collection of various multichannel stereophonic microphone techniques, overlaid on top of each other, each used where appropriate. Other complex productions, like live sporting events, may use some of the same techniques. For instance, it is commonplace to use spaced microphones for ambience of the crowd, supplying their outputs to the surround channels, although it is also important, at one and the same time, not to lose intimacy with the ongoing event. That is, the front channels should contain "close up" sound so that the ambience does not overwhelm the listener. This could be done with a basic stereo pickup for the event, highlighted by spot mikes.

For instance, years ago one Olympic downhill skiing event employed French Army personnel located periodically along the slope. Each was equipped with a shotgun microphone, and was told to use it like a rifle, aiming at the downhill skier. Crossfading from microphone to microphone kept the live sound mixer on his toes, but it also probably resulted in the most realistic perspective of involvement with the event that was ever heard. Contrast that with more recent Olympic gymnastic events where what you hear is the crowd in a large reverberant space while what you see is a close up of feet landing on a horse, which you can barely hear if at all in the sea of noise and reverberation. Microphone technique dominates these two examples.

Some Surround Microphone Setups

Some surround microphone setups are listed in Table 3-2. While these recordings were made for the 10.2-channel system, many of the features apply to 5.1, 7.1, and other systems.

Simultaneous 2- and 5-Channel Recording

John Eargle has developed a method that combines several microphone techniques and permits the simultaneous recording of 2-channel stereo for release as a CD and the elements needed for 5.1-channel release. This technique is used so the recording has "forward compatibility" for the future since the marketplace for classical music is driven by 2-channel CD sales, so recognizes the importance of that market, while providing for a future market. The process also allows for increased word length to 20 bits through a bit splitting recording scheme, all on one 8-track digital recorder that is easily portable to the venues used for recording such music.

The technique involves the following steps:

- At the live session, record to digital 8-track the following:
  1. Left stereo mix
  2. Right stereo mix
  3. Left main mic., LM
  4. Right main mic., RM
  5. Left house mic.
  6. Right house mic.
  7. Bit splitting recording for 20 bits
  8. Bit splitting recording for 20 bits

$$\text{Left stereo mix} = \text{LM} + (g)\,\text{LF}$$

$$\text{Right stereo mix} = \text{RM} + (g)\,\text{RF}$$

  where (g) is the relative gain of the outrigger mic. pair compared to the main stereo pair, such as −5 dB, and LF and RF are left and right front outrigger mics.

Table 3-2 Microphone Usage on One Orchestral Recording of Classical Music

Program	Microphone setup	Microphone no. and type	Notes
New World Symphony: Copeland Symphony No. 2, Fanfare for the Common Man	Main ORTF Pair, Cardioid	2 × Sanken CU-41 over head and behind conductor	Pan just to the left and right of center
	Secondary Main Pair, Sphere	1 × Schoeps Sphere behind main pair	Used for test only
	L/R Outriggers, Omni	2 × Schoeps MK2* at 1/4 and 3/4 of the width of the orchestra in line with the main pair	
	Spot Mikes, Omni	11 × Schoeps MK2: 2 × harps, 2 × tymps, 1 × kettle drum, 2 × other percussion, 2 × woodwinds, 2 × double bass	Kettle drum mike with screw-in pad for 138 dB SPL peak level in close miking described below
	Surround Sphere	1 × Schoeps Sphere plus bidirectional pair with M-S matrix decoding	The hall in which this recording was made was rather poor for surround recording, so in the end artificial reverberation was used instead of these microphone channels
	Surround Microphones, backwards-facing cardioid in balcony	2 × Neumann TLM 103	
	Supply air duct	1 × omni electret covered in condom	Experimental purpose to see if in-room noise could be reduced through correlation with supply air duct noise
Shakespeare *Merry Wives of Windsor* theater in the round, inside out	Spaced omnis on a 10-foot diameter circle with angles corresponding to loudspeakers: 0°, ±30°, ±60°, ±110°, 180°	8 × Schoeps MK2	Audience meant to be in center; play staged all round
	Spot mikes on musical instruments	2 × Sennheiser MKH40	
USC Marching Band at Homecoming	Spaced omnis on a line straddling the 50-yard line and up against the stands spaced a total of 60 ft for five microphones	5 × Schoeps MK2	With screw-in pre-electronics −10 dB pads as there was no rehearsal and hundreds of players on the field
	Spaced omnis on the same line further downfield in front of unoccupied seats	3 × Schoeps MK2	Used in the end for surround as they provided a more distant perspective
	Dummy head in stands with people around it	1 × KEMAR (2 channels)	Used for surround close-up perspective with equalization for ear-canal resonance
Herbie Hancock's tune "Butterfly"	48-track studio recording close miked, overdubbed	Unknown, but close miked and direct box outputs	Mix theory covered in next chapter.

* For those microphones that are modular only the pickup capsule is listed. Corresponding electronics are also used of course.

- Use left and right stereo mix for 2-channel release. In a postproduction step, subtract left and right main mikes from the stereo mix to produce left and right tracks of a new LCR mix, and add the main mikes back in, panning them into position using two stereo pan pots between center and the extremes. In order to do a subtraction, set the console gain of the left and right main mics. to the same setting (g) as used in the original recording, and flip the phase of the left and right main mic. channels using console phase switches, or with balanced cables built that reverse pins 2 and 3 of the XLRs or equivalent in those channel inputs.
- For the final 5.1 mix, use the LCR outputs of the console shown in the figure, and the house mics. for left and right surround. Most classical music does not require 0.1 channel because flat headroom across frequency is adequate for the content (see Fig. 3-10 for a diagram of this method).

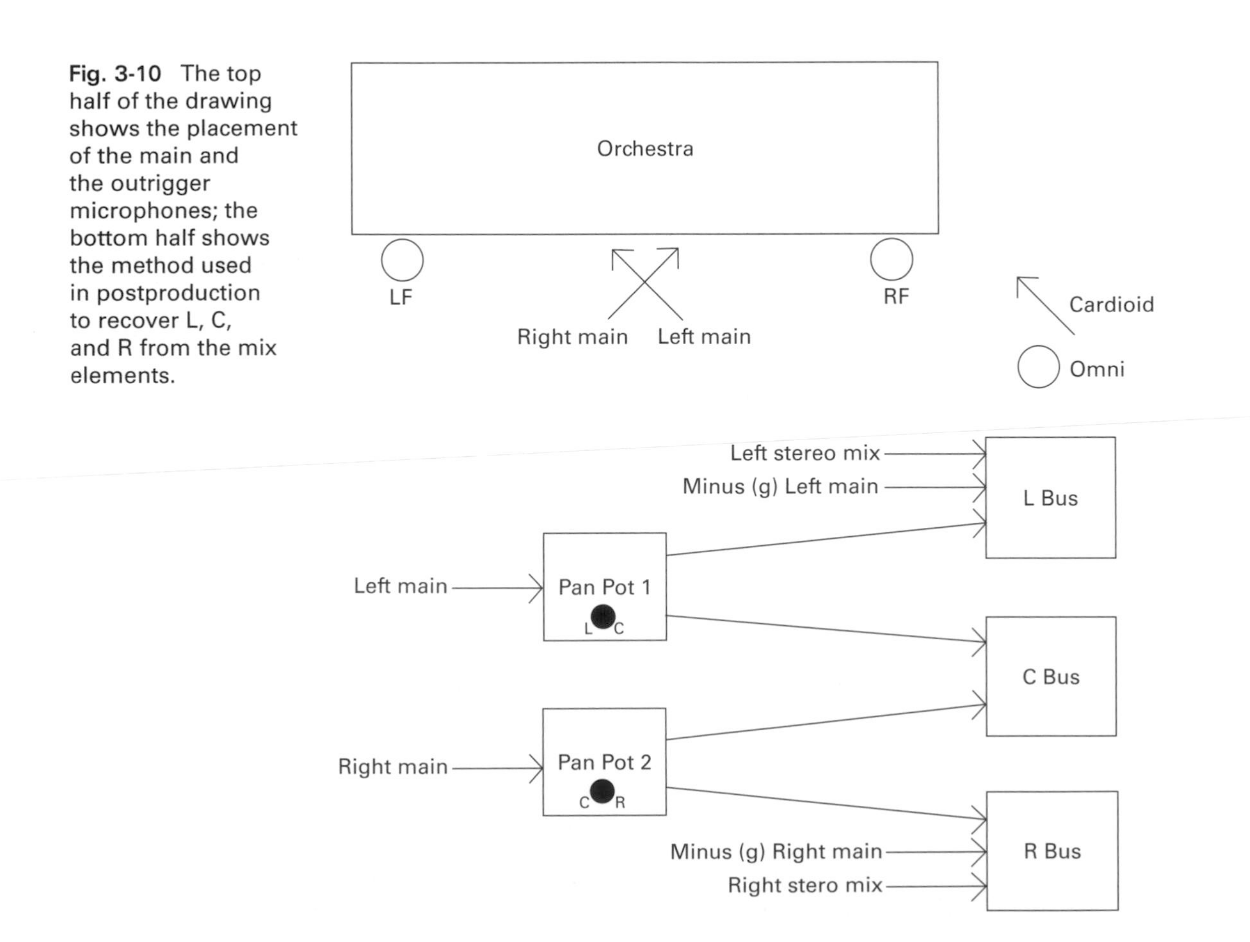

Fig. 3-10 The top half of the drawing shows the placement of the main and the outrigger microphones; the bottom half shows the method used in postproduction to recover L, C, and R from the mix elements.

Upmixing Stereo to Surround

While all-electronic methods may be used to upmix existing 2-channel program material to surround, such as adding reverberation from an electronic reverberator, use of a good sounding room to derive LS and RS channels from left/right stereo mixes is also known to be useful. Use of backwards-facing cardioids or a Hamasaki square is most useful, and the loudspeakers driving the room must be of high quality, with smooth power response since it is the reverberant field of the room that dominates in the microphones outputs. Note that surround recordings can often have more relative reverberation than stereo ones since the reverberation is spread out among more channels, and thus does not tend to clutter up and make less distinct the front channels, as it would in 2-channel stereo.

Center channel derivation is more difficult. Simply summing L+R and sending it attenuated to center does little to improve mixes typically, and often may cause problems such as a narrowing of the stereo sound field. More sophisticated extraction of center is available from certain processors on the market.

Dynamic Range: Pads and Calculations

For most recording, it is possible to rehearse and set a level using the input trim control of the mic. preamp for the best tradeoff between headroom and noise. What if you don't get a rehearsal? Maybe then you use your experience in a given situation (instrument(s), room, microphone, gain structure). But if it's a "cold" recording, what then? Well then you can calculate the required preamp gain from the expected sound pressure level, the microphone sensitivity, and the input sensitivity of the device being fed, such as a recorder.

The first thing to consider is whether the expected sound pressure level at the microphone might clip the microphone's own electronics in the case of electrostatic microphones, those most often used for this type of recording. For the New World Symphony recording spot mikes described in the table above I used a memory of measurement of another bass drum from another time. I measured the inside of a bass drum head of REO Speedwagon in my studio in the environs of Champaign, IL in the late 1960's by putting a Shure SM-57 dynamic mike inside the head and feeding it into an oscilloscope directly. By calibrating the scale, I was able to find the peak sound pressure level of 138 dB.

The Schoeps electronics operating on 48 V phantom clip at just over 130 dB, but they don't reach 138 dB. So we needed pads, ones that screw in between the capsule and the electronics. We used 10 dB pads on the five percussion spot mics. as these were the ones that were close enough to instruments loud enough to cause potential problems, and the signal-to-noise ratio was not harmed as the instruments were so loud that their spot mic. contribution was well down in the mix.

Then to calculate the gain, we use the microphone sensitivity, the pad value, and the input sensitivity of the recorder for Full Scale to determine the unknown in the overall equation—the gain setting of the microphone preamplifier. The mic. sensitivity is 15 mV/Pa, the pad is 10 dB, and the input sensitivity of the recorder used for Full Scale is 18 dB over +4 dBu, namely +22 dBu. Let's take 138 dB SPLpk as our level that must be handled cleanly. Then rms level is 135 dB, and the rest of our calculation can be in rms. With 15 mV at 1 Pa which equals 94 dB SPL, 135 dB SPL is 41 dB hotter. The 10 dB pad takes this down to 31 dB hotter. 31 dB up from 15 mV is 530 mV (calculated by dividing 31 by 20, then raising 10 to the power of the remainder: 10^(31/20) = 35.5 times. 15 mV × 35.5 = 530 mVrms. +22 dBu = 9.76Vrms. (Get from 10^(22/20) × 0.775 (the reference level for 0 dBu) = 9.76Vrms.) Now 20 log (9.76/0.53) = 25 dB). So the maximum gain we can use is 25 dB, and to leave a little headroom beyond 135 dB SPL in case this drum is louder, let's make it 20 dB. We did. It worked. The maximum recorded level hit about −3 dBFS, and with 24-bit recording, we had a huge dynamic range captured.

This can be done with a $10 Radio Shack scientific calculator and 5 minutes, by following the equations above. Or it's more fun to do it in your head and amaze your friends. The trick is to know how to do dB in your head by learning a few numbers: 1 dB = 1.1, 3 dB = 1.4, 6 dB = 2, 10 dB = 3.1, 20 dB = 10. From this you can decompose any number of decibels quickly. Take 87 dB; 80 dB is a factor of 10,000 (from four twenties); 7 dB more is a factor of two times 1.1 (6 + 1 dB). So 87 dB is 22,000, in round numbers.

We also performed a back of the napkin calculation during breakfast at USC Homecoming some years ago. We were to record the Marching Band, and all its alumni available were asked to play too, so it was quite a group. While our microphones were not on top of them, still the sound power of a full marching band probably doubled in numbers with alums is quite stunning. I was on the sidelines as boom operator with a spot mike (which we didn't use in the end) and it was so loud you couldn't communicate with each other (and boy did I have my earplugs in!). No rehearsal. We calculated the mike gain (the preamps we were using had step gain controls, not easy to change smoothly at all during recording) on the napkin over breakfast and were very happy when the peak levels also hit about −3 dBFS. So this method, that seems a little technical to most people in our industry, shows the utility of doing a little math, of estimating, of using decibels well, and so forth. When you are bored in math class just think of the loud but exceptionally clean recordings you'll be able to make when you master this skill; nobody told me that at the time!

Virtual Microphones

My colleague at USC Chris Kyriakakis has studied the recordings that we have made in 10.2 and come up with a method of upmixing existing 2-channel recordings to an arbitrary number of channels with microphone positions that never existed on the original session. Use is made of the differences between the main stereo pair, and other microphones like hall mikes of our 10.2-channel recordings. By use of high-order mathematics, the difference between the two microphones in 3-D space may be found and applied to other different recordings that are dominated by main microphones. Thus hall mikes, and even spot mikes, not present in the original may be provided to the producer/engineer for new multichannel mixes.

4 Multichannel Mixing and Studio Operations

Tips from This Chapter

- Surround sound allows for enveloping (surrounding) the listener, not just spaciousness.
- Two basic approaches exist for front/surround sound: direct/ambient and direct-sound all around.
- Many microphone channels can be hard assigned to loudspeaker channels, while some need fixed panning in between two adjacent channels, and some may need the ability to pan dynamically.
- Panners come in various forms: three knobs, joystick, and DAW software. Panning laws, divergence, and focus controls are explained in the text. Work arounds for 2-channel oriented equipment are given.
- Panning is the greatest aesthetic frontier of multichannel sound. One major error is panning sources between left and right, ignoring center.
- Source size can be changed through signal processing, including surround decoding 2-channel sources, decorrelation, and reverberation.
- Equalizing for multichannel recordings varies from stereo: centered content will probably use less equalization, while direct sound sources placed in the surrounds may require more equalization than typical, in order to maintain the timbre of the source.
- Multichannel routing is usually by way of three AES-3 standard pairs, organized according to the track format in use, of which there are several. Jitter can affect reception of digital audio signals, and digital audio receivers must be well designed to reject the effects of jitter.
- The most common multichannel track assignments is: L, R, C, LFE, LS, RS. Several other track assignments exist.

- Multichannel audio for video is delivered by means of a special bit-rate reduction method called mezzanine coding so that it may be stored on the available digital audio tracks of video tape recorders, or by means of a digital audiotape separate from the video (double system) because most video recorders only have 4 channels. Time code must match the picture, and 48 kHz is the standard sample rate.
- Mezzanine (lighter than transmission-coded audio) coding makes two of the digital videotape audio channels capable of carrying 5.1 audio channels, with multiple generation coding/decoding cycles possible and with editing features not found in the transmission codec. One such codec is Dolby E.
- SMPTE reference level is –20 dBFS; EBU is –18 dBFS.
- Multichannel monitoring electronics must include source-playback switching, solo/mute/dim, and monitor volume control, with calibration, individual level trims, bass management, and method of checking mixdown to stereo and mono.
- Outboard dynamics units (compressors, limiters, etc.) need multichannel control links.
- Several reverberators with stereo outputs can be used to substitute for one reverberator with five outputs.
- Decorrelators are potentially useful devices, including those based on pitch shifting, chorus effect, and complementary comb filters.

Introduction

At the start of the multichannel era for music, several pop music engineer–producers answered a question the same way independently. When asked, “How much harder is it to mix for 5.1-channel sound than stereo?” they all said that 5.1-channel mixes are actually easier to perform than 2-channel ones. This surprised those whom had never worked on multichannel mixes, but there is an explanation. When you are trying to “render” the most complete sonic picture, you want to be able to distinguish all of the parts. Attention to one “stream” of the audio, such as the bass guitar part, should result in a continuous performance on which you may concentrate. If we then subsequently pay attention to a lead vocal, we should be able to hear all the words. It is this multistream perceptual ability that producers seek to stimulate (and which keeps musical performances fresh and interesting, despite many auditions). By mixing down to 2 channels, first the bass and vocal need compression, then equalizing. This is an interactive process done for each of the tracks, basically so that the performances are each heard without too much mutual interference. The bottom line on 2-channel stereo is that there is a small box in which to put the program content, so each part has to be carefully tailored to show through the medium.

A great illustration of this effect is the Herbie Hancock album *Butterfly.* This complex work was written and produced with surround sound in mind, but since there was no simple delivery means for it in the era in which it was made, the record company asked for a 2-channel version. Upon mixing it down to 2 channels, the complexity of the various musical lines was no longer clear, despite bringing all the resources of a high-end 2-channel stereo mix to bear. The album failed in the marketplace, probably because the producers just tried to pack too much sound into too few channels.

With more channels operating, there is greater likelihood that multiple streams can be followed simultaneously. This was learned by the US Army in World War II. Workers in command and control centers that were equipped with audible alerts like bells, sirens, and klaxons, perceived the separate sources better when they were placed at multiple positions around the control room, rather than when piled up in one place. This "multi-point mono" approach helps listeners differentiate among the various sources. Thus, you may find that mixing multichannel for the first time is actually easier than it might seem at first glance. Sure, the mechanics are a little more difficult because of the number of channels, but the actual mixing task may be easier than you thought.

Multi-point mono, by itself, is not true stereo because each of the component parts lacks a recorded space for it to "live" in. The definition of the word stereo is "solid," that is, each sound source is meant to produce a sensation that it exists in a real 3-D space. Each source in an actual sound field of a room generates three sound fields: direct sound, reflected sound, and reverberation. Recording attempts to mimic this complex process over the facilities at hand, especially the number of channels. In 2-channel stereo, it is routine to close mike and then add reverberation to the mix to "spatialize" the sound. The 2-channel reverberation does indeed add spaciousness to a sound field, but that spaciousness is largely constrained to the space between the loudspeakers. What 2-channel stereo lacks is another significant component of the reproduction of space: envelopment, the sense of being immersed in and surrounded by a sound field. So spaciousness is like looking into a window that contains a space beyond; envelopment is like being in the space of the recording. What multichannel recording and reproduction permits is a much closer spatial approximation to reproducing all three sound fields of a room than can 2 channels. In this chapter we take up such ideas, and give specific guidelines to producing for the multichannel medium.

Mechanics

Optimally, consoles and monitor systems should be designed for at least the number of loudspeaker channels to be supported, typically 5.1 (i.e., six electrical paths), and it is commonplace to design digital audio workstations (DAW), recorders, and consoles in 8-channel groups. The principal parts of console design affected by multichannel are the panning and subsequent output assignment and bussing of each of the input channels, and the monitoring section of the console, routing the console output channels to the monitor loudspeaker systems. Complete console design is beyond the scope of this book, but the differences in concepts between stereo consoles and multichannel ones will be given in some detail.

Panners

Multichannel panning capability is a primary difference between stereo, on the one hand, and 5 channels and up consoles on the other. Although multichannel consoles typically provide panning on each input channel, in fact many of the input channels are assigned directly to just one output channel and wind up in one loudspeaker. This may be exploited in multibus consoles that lack multichannel panning, since a lot of panning is actually hard channel assignment. An existing console can thus be pressed into service, so long as it has enough busses. Input channels that need dynamic panning among the output channels may be equipped with outboard panners, and outboard monitoring systems may be used.

There are three basic forms of multichannel panners. The principal form that appears in each channel of large-format consoles uses three knobs: left–center–right, front–back, and left surround–right surround. This system is easier to use than it sounds because many pans can be accomplished by presetting two of the knobs and performing the pan on the third knob. For instance, let's say that I want to pan a sound from left front to right surround. I would preset the L/C/R knob to left, and the LS/RS knob to RS, and perform the pan on the F/S knob when it is needed in the mix (Fig. 4-1).

The advantage of this panner type over the next to be described is that the cardinal points, those at the loudspeaker locations, are precise and emphasized because they are at the extremes of the knobs, or at a detent (click) provided to indicate the center. It is often preferred to produce direct sound from just one loudspeaker rather than two, because sound from two produces phantom images that are subject to the precedence effect, among other problems described in Chapter 6.

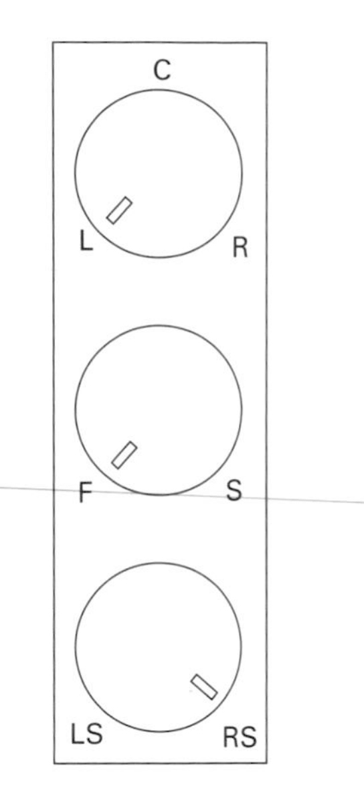

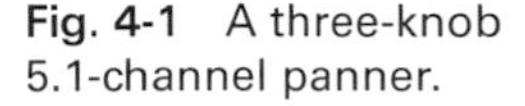
Fig. 4-1 A three-knob 5.1-channel panner.

The precedence effect, or Law of the First Wavefront, says that we localize sound to the first arriving location, so if sound is panned halfway between left and center and we are sitting to the left of center, then the sound location will be distorted towards the left loudspeaker.

The second type of panner is the joystick. Here, a single computer game-style controller can move a sound around the space. This type emphasizes easier movement, at the expense of precision in knowing where the sound is being panned. It also emphasizes the "internal" parts of the sound field, where the source is sent to L, C, R, LS, and RS all simultaneously, for that is what such panners will typically do when set with the joystick straight up. This is often not a desirable situation since each listener around a space hears the first arriving direction—so a huge variety of directions will be heard depending exactly on where one is sitting, and a listener seated precisely at the center hears a mess, with each loudspeaker's sound affected by the associated head-related transfer functions (HRTFs). What is heard by a perfectly centered listener to perfectly matched loudspeakers driven together, is different frequency regions from different directions, corresponding to peaks in the HRTFs. It sounds as though the source "tears itself apart" spectrally.

Upon panning from front to surround, certain frequency ranges seem to move at first, then others, only to come back together as the pan approaches one single channel. Thus, although it may seem at first glance that joystick-based panning would be the most desirable from the standpoint of ease of use, in fact, most large-format consoles employ the three-knob approach, not simply because it fits within the physical constraints of a channel slice, but because the emphasis of the three-knob panner is more correct.

The third type of multichannel panner is software for DAW. Various "plug-ins" are available for multichannel panning, and it is only a matter of time before multichannel panning is a core feature of DAW software. Advantages of software panners include automation and the potential linking of channels together to make a pair of source channels "chase" each other around a multichannel space. This is valuable because practically all sound effects recordings are 2-channel today, and it is often desirable to spatialize them further into 5.1. Methods for doing this will be described below.

Work Arounds for Panning with 2-Channel Oriented Equipment

Even when the simple "hard" assignment of input to output channels needs some expansion to panning of one input source "in between" two output channels, still only a 2-channel panner is necessary, with the outputs of the panner routed to the correct 2 channels. So interestingly, a console designed for multitrack music production may well have enough facilities for straightforward multichannel mixes, since panning for many kinds of program material is limited to a hard channel assignment or assignment "in between" just 2 channels. For such cases, what the console must have is an adequate number of busses and means of routing them to outputs for recording and monitoring. These could be in the form of main busses or of auxiliary busses. Thus, a console with a stereo 2-channel mixdown bus and with 4 aux busses can be pressed into 5.1-channel service (although you have to keep your wits about you).

Clearly, it is simpler to use purpose-built multichannel equipment than pressing 2-channel equipment into multichannel use, but it is worth pointing out that multichannel mixes can be done on a 2-channel console with adequate aux sends. With this feature an adequate number of pairs of channels can be represented and thus input channels mapped into pairs of output channels, and "pair-wise" panning performed. Here is how this is done:

- Use output channel assignments for the medium in use. For 8-track multichannel mixes for television, this is seen later as: (1) left, (2) right, (3) center, (4) LFE, (5) left surround, (6) right surround.
- Assign an input track such as 1, to a bus pair, such as bus 1–2.
- Assign the bus pair to output channel pairs, bus 1 and 2 to output channels 1 and 3, respectively. (This requires that you build aux input tracks.)
- Now the stereo panner on input track 1 pans between left and center.

- If you have to continue a moving pan from left through center to right, then you will have to split the track in two at the point where it goes through center, using the first track for the first part of the pan, and the second track for the second part. This is because there are only 2-channel panners and no dynamic bussing during a session. Although this is clumsy, it does work.

Panning Law

The "law" of a control is the change of its parameters with respect to mechanical input. For a volume control, this is represented by the scale next to or around the control knob that shows you the attenuation in decibels for different settings of the control. For a panner, at least two things are going on at once: one channel is being attenuated while another is fading up as the control is moved. The law of a panner is usually stated in terms of how many decibels down the control is at its midpoint between 2 channels. Very early work at Disney in the 1930s determined that a "power law" was best for panners, wherein the attenuation of each of the channels is 3 dB at the crossover point. This works perfectly in large reverberant environments like the original Disney dubbing stages because sound adds as sound power in the reverberant field, and two sources 3 dB down from one source will produce the same sound pressure level as the single source. However, things get more complicated when the sound field is mixed among direct, reflected, and reverberant. The BBC found in the 1970s that about 4.5 dB down in each of 2 channels produced equal level perception as a single source when sound was panned between channels in an environment much more like studio control rooms and home listening ones. The 3 dB down law is also called a "sin–cos" function, because the attenuation of 1 channel, and the increasing level of another, follow the relationship between the sine and cosine mathematical functions as the knob is turned.

In fact, panning based simply on level variation among channels greatly simplifies the actual psychoacoustics of what is going on with a real source. Amplitude panning works best across the front, and again across the back, of a 5.1-channel setup, but works poorly on the sides, for reasons explained in Chapter 6.

An additional knob on some panners is called divergence. Divergence controls progressively "turn up" the level in the channels other than the one being panned to (which is at full level), in order to provide a "bigger" source sound. With full divergence, the same signal is sent to all of the output channels. Unfortunately, sounds panned with divergence are subject to the precedence effect, and putting the same sound into all of the loudspeaker channels causes the listener to locate the sound to the closest loudspeaker to their listening position, and the production of highly audible timbre and image shift effects.

It is an interesting historical note that the invention of divergence came in the early 1950s in Hollywood, about the same time that Helmut Haas was finding the summing localization effect called Law of the First Wavefront described in Chapter 6. It seems highly unlikely that the inventors in Hollywood were reading German journals, so they had probably never heard of this effect, and did not know the consequences of their actions. The motivation was also not only due to the idea of getting a bigger sounding source, but also to reduce the obvious timbre variations as an actor would enter screen left and exit screen right panning across up to 5 channels, with a different timbre produced by each channel due to the lack of room equalization and tolerances of loudspeaker drivers in these early systems. Divergence helped to conceal the major timbre shifts.

It should be said however that the curiosity of sounding comb filtered and of localizing to the nearest loudspeaker has been put to good use at least once. In the voice overs of the main character in *Apocalypse Now*, the intimate sounding narration is piped to all three front channels—the divergence control would be fully up at least with respect to the front channels on the console. This helps, along with the close-miked recording method, to distinguish the "inside the head" effect of voice over, and lends maximum intimacy to all parts of the cinema because as one moves left and right across a row one finds the voice over to stay continuously more or less in front of one. That is, as you move off the centerline the narrator's image shifts until you are in front of the left channel loudspeaker, and straight in front of you now is the direction you hear. There is certainly some comb filtering involved in driving all three front loudspeakers, but in this case it is not seen as a defect because the more different the voice over sounds from production dialogue the better. Walter Murch, sound designer of the picture, has talked about the fact that the narration was recorded three times—for script reasons, for performance reasons, and for recording method reasons—with improvements each time. He has also said that he took as a model the voice over in the earlier film *Shane* to distinguish voice over from on-screen dialogue.

A further development of the divergence concept is the focus control. Focus is basically "divergence with shoulders." In other words, when sound is panned center and the focus control is advanced off zero, first sound is added to left and right, and then at a lower level, to the surrounds. As the sound is panned, the focus control maintains the relationship; panned hard right, the sound is attenuated by one amount in center and right surround, and by a greater amount in left and left surround. Focus in this way can be seen as a way to eliminate the worst offenses of the divergence control. If a source needs to sound larger, however, there are other methods described below.

The Art of Panning

Panning is used in two senses: fixed assignment of microphone channels to one or more loudspeaker channels, called static panning, and motion of sound sources during mixing, called dynamic panning. Of the two, static panning is practiced on nearly every channel every day, while dynamic panning is practiced for most program material much less frequently, if at all.

The first decision to make regarding panning is the perspective with which to make a recording: direct/ambient or direct-sound all round. The direct/ambient approach seeks to produce a sound field that is perceived as "being there" at an event occurring largely in front of you, with environmental sounds such as reverberation, ambience, and applause reproduced around you. Microphone techniques described in Chapter 3 are often used, and panning of each microphone channel is usually constrained to one loudspeaker position, or in between two loudspeaker channels. The "direct" microphones are panned across the front stereo stage, and the "ambient" microphones are panned to the surround channels, or if there are enough of them to the left/right and left surround/right surround channels. Dynamic panning would be unusual for a direct/ambient recording, although it is possible that some moving sources could be simulated.

The second method, called "direct-sound all round," uses the "direct" microphone channels assigned to, typically, any one or two of the loudspeaker channels. Thus, sources are placed all around you as a listener—a "middle of the band" perspective. For music-only program, the major aesthetic question to answer is, "What instruments can be placed outside the front stereo stage and still make sense?" Instruments panned part way between front and surround channels are subject to image instability and sounding split in two spectrally, so this is not generally a good position to use for primary sources, as shown in Chapter 6. Positions at and between the surround loudspeakers are better in terms of stability of imaging (small head motions will not clearly dislodge the sound image from its position) than between front and surrounds.

Various pan positions cause varying frequency response with angle, even with matched loudspeakers. This occurs due to the HRTFs: the frequency response occurring due to the presence of your head in the sound field measured at various angles. While we are used to the timbre of instruments that we are facing due to our conditioning to the HRTF of frontal sound, the same instruments away from the front will demonstrate different frequency response. For instance, playing an instrument from the side will sound brighter compared to in front of you due to the straight path down your ear canal of the side location. While some of this effect causes localization at the correct positions and thus can be said to be a part of natural listening, for careful listeners the effect on timbre is noticeable.

The outcome of this discussion is simply this: you should not be afraid to equalize an instrument panned outside the front stereo stage so that it sounds good, rather than thinking you must use no equalization to be true to the source. While simple thinking could apply inverse HRTF responses to improve the sound timbre all round, in practice this may not work well because each loudspeaker channel produces direct sound subject to one HRTF for one listener position, but also reflected sound and reverberation subject to quite different HRTFs. Thus, in practice the situation is complex enough that a subjective view is best, with good taste applied to equalizing instruments placed around the surround sound field.

In either case, direct/ambient or direct-sound all round, ambient microphones picking up primarily room sound are fed to the surround channels, or the front and the surround channels. It is important to have enough ambient microphone sources so that a full field can be represented—if just two microphones are pressed into service to create all of the enveloping sound, panning them halfway between front and surround will not produce an adequate sense of envelopment. Even though each microphone source in this case is spacious sounding due to their being reverberant, each one is nonetheless mono, so multiple sources are desirable.

In the end, deciding what to pan where is the greatest aesthetic frontier associated with multichannel sound. Perhaps after a period of experimentation, some rules will emerge that will help to solidify the new medium for music. In the meantime, certain aesthetic ideas have emerged for use of sound accompanying a picture:

- The surround channels are reserved typically for reverberation and enveloping ambience, not "hard effects" that tend to draw attention away from the picture and indicate a failure of completeness in the sensation of picture and sound. Called the exit sign effect, drawing attention to the surrounds breaks the suspension of disbelief and brings the listener "down to earth"—their environs, rather than the space made by the entertainment.

- Certain hard effects can break the rule, so long as they are transient in nature. A "fly by" to or from the screen is an example.

Even with the all round approach, most input channels will likely be panned to one fixed location for a given piece of program material. Dynamic panning is still unusual, and can be used to great effect as it is a new sensation to add once a certain perspective has been established.

Non-Standard Panning

Standard amplitude panning has advantages and disadvantages. It is conceptually simple, and it definitely works over a large audience area at the "cardinal" points, that is, if a sound is panned hard left all audience members will perceive it as at the left loudspeaker. Panning halfway between channels leads to some problems, as moving around the listening area will cause differing directional impressions, due to the precedence effect described in Chapter 6. Beyond conventional amplitude panning are two variations that may offer good benefits in particular situations. The first of these is time-based panning. If the time of arrival of sound from two loudspeakers is adjusted, panning will be accomplished having similar properties to amplitude panning. Second, more or less complete HRTFs can be used in panning algorithms to better mimic the actual facts of a single source reproduced in between two loudspeakers. If a sound is panned halfway between left and left surround loudspeakers, it is often perceived as breaking up into two different events having different timbre, because the frequency response in the listener's ear canal is different for the two directions of arrival. By applying frequency and time response corrections to each of the two contributory channels it is possible for a sound panned between the two to have better imaging and frequency response. The utility of this method is limited in the listening area over which it works well due to the requirement for each of the channels to have matching amplitude and time responses. One console manufacturer employs HRTF and time-based panning algorithms and room simulation within the console and that is Studer.

Panning in Live Presentations

When the direct/ambient approach is used for programming such as television sports, it is quite possible to overdo the surround ambience with crowd noise, to the detriment of intimacy with the source. Since the newly added sensation the last few years is surround, it is the new item to exercise, so may be overused. What should not be forgotten is the requirement for the front channels to contain intimate sound. For instance, in gymnastics, the experience of "being there" sonically is basically hearing the crowd around you, with little or no sound from the floor. But television is a close-up medium, and close-ups, accompanied by ambient hall sound, are together a disjointed presentation. What is needed is not only the crowd sounds panned to the surrounds, but intimate sound in front. The crowd should probably appear in all the channels, decorrelated by being picked up by multiple microphones. Added to this should be close-up sound, probably shotgun based, usually panned to center, showing us the struggle of the gymnasts, including their utterances, and the squeak and squawk of their interfacing with the equipment. In this case, the sense of a stereo space is provided by the ambient bed of crowd noise, helping to conceal the fact that the basic pickup of the gymnasts is in fact mono. In a live event, it is improbable that screen direction of left–center–right effects can be tracked quickly enough to make sense, so the mono spot mic. plus multichannel ambience is the right combination of complexity (namely, simple to do) and sophistication (namely, sounds decent) to work well.

A Major Panning Error

One error that is commonplace treats the left and right front loudspeakers as a pair of channels with sound to be panned between them—with the center treated as extra or special. This stems from the thinking that the center channel in films is the "dialogue channel," which is not true. The center channel, although often carrying most if not all of the dialogue, is treated exactly equal to the left and right channels in film and entertainment television mixes, for elements ranging from sound effects through music. It is a full-fledged channel, despite the perhaps lower than desired quality of some home theater system center loudspeakers.

What should be done is to treat the center just as left and right. Pans should start on left, proceed through center, and wind up at right. For dynamic pans, this calls for a real multichannel panner. Possible work arounds include the method described above to perform on DAWs: swapping the channels at the center by editing so that pans can meet the requirement. What panning elements from left to right and ignoring center does is to render the center part of the sound field so generated as a phantom image, subject to image-pulling from the precedence effect, and frequency response anomalies due to two loudspeakers creating the sound field meant to come from only one source as described in Chapter 6.

As of this writing when I attend movies and sit on the centerline you will find me during the end credit music leaning left and right to see for sure that the music has been laid in 2-track format. Live sound situations too often rely on 2-channel stereo left and right since center runs into problems with other things there, like the band, and flying clusters are difficult. However, 2-channel stereo works worse in large spaces than in small ones because the time frame is so much larger in the big space. Given that you can hear an error of 10μs(!) in imaging of a center front phantom, in fact there is virtually no good seating area in the house, but rather only along a line perpendicular to the stage on the centerline. This is why I always get to movies early, or in Los Angeles buy a ticket in advance to a specific seat, so I can sit on the centerline and get the best performance! While it probably is cheaper to take existing 2-channel music and lay it in rather than process it to extract a center, the consequence is that only a tiny fraction of the audience hears the music properly centered.

Increasing the "Size" of a Source

Often the apparent size of a source needs to be increased. The source may be mono, or more likely 2-channel stereo, and the desire exists to expand the source to a 5-channel environment. There is a straightforward way to expand sources from 2 to 5 channels:

- A Dolby SDU-4 (analog) or 564 (digital) surround sound decoder can be employed to spatialize the sound into at least 4 channels, L/C/R/S (LS and RS being the same).
- Of course, it is possible to return the LCRS outputs of the surround sound decoder to other channels, putting the principal image of a source anywhere in the stereo sound field, and the accompanying audio content into adjacent channels. Thus, LCRS could be mapped to CRSL, if the primary sound was expected to be in right channel.

For the 2:5 channel case, if the solution above has been tried and the sound source is still too monaural sounding after spatialization with a surround decoder, then the source channels are probably too coherent, that is, too similar to one another. There are several ways to expand from 1 channel to 5, or from 2 channels that are very similar (highly correlated) to 5:

- A spatialization technique for such cases is to use complementary comb filters between the 2 channels. (With a monaural source, two complementary combs produce two output channels, while with a 2-channel source, adding complementary comb filters will make the sound more spacious.) The response of the 2 channels adds back to flat for correlated sound, so mixdown to fewer channels remains good. The two output channels of this process can be further spatialized by the surround sound decoder technique. Stereo synthesizers intended for broadcasters can be used to perform this task, although they vary from pretty awful to pretty good depending on the model.

- One way to decorrelate useful for sound effects is to use a slight pitch shift, on the order of 5–10 cents of shift, between two outputs. One channel may be shifted down while the other is shifted up. This technique is limited to non-tonal sounds, since strong tones will reveal the pitch shift. Alternatives to pitch shift-based decorrelation include the chorus effects available on many digital signal processing boxes, and time-varying algorithms.
- Another method of size changing is to use reverberation to place the sound in a space appropriate to its size. For this, reverberators with more than two outputs are desirable, such as the Lexicon 960. If you do not have such a device, one substitute is to use two stereo reverberators and set the reverberation controls slightly differently so they will produce outputs that are decorrelated from each other. The returns of reverberation may appear just in front, indicating that you are looking through a window frame composed of the front channels, or they may include the surrounds, indicating that the listener is placed in the space of the recording. Movies use reverberation variously from scene to scene, sometimes incorporating the surrounds and sometimes not. Where added involvement is desired, it is more likely that reverberation will appear in the surrounds.
- For reverberators with four separate decorrelated outputs the reverb returns may be directed to left, right, left surround, and right surround, neglecting center. Center reverberation, particularly of dialogue, tends to be masked by direct sound and so is least effective there.

Equalizing Multichannel

The lessons of equalizing for stereo apply mostly to multichannel mixing, with a few exceptions noted here.

Equalizing signals sent to an actual center channel is different from equalizing signals sent to a phantom center. For reasons explained in Chapter 6, phantom image centered stereo has a frequency response dip centered on 2 kHz, and ripples in the response at higher frequencies. This dip in the critical mid-range is in the "presence" region, and it is often corrected through equalization, or through choice of a microphone with a presence peak. Thus it is worth it not to try to copy standard practice for stereo in this area. The use of flatter frequency response microphones, and less equalization, is the likely outcome for centered content reproduced over a center channel loudspeaker.

As described above, and expanded in Chapter 6, sound originating at the surrounds is subject to having a different timbre than sound from the front, even with perfectly matched loudspeakers, due to HRTF effects. Thus, in the sound-all-round approach, for sources panned to or between the surround loudspeakers, extra equalization may be necessary to get the timbre to sound true to the source. One possible equalization to try is given in Chapter 6.

In direct/ambient presentation of concert hall music the high frequency response of the surround channel microphones is likely to be rolled off due to air absorption and reverberation effects. It may be necessary to adjust any natural recorded rolloff. If, for instance, the surround microphones are fairly close to an orchestra but faced away, the high-frequency content may be too great and require roll-off to sound natural. Further, many recording microphones "roll up" the high frequency response to overcome a variety of roll offs normally encountered, and that is not desirable when used in this service.

Routing Multichannel in the Console and Studio

On purpose-built multichannel equipment, five or more source channels are routed to multichannel mixdown busses through multichannel panners as described above. One consideration in the design of such consoles is the actual number of busses to have available for multichannel purposes. While 5.1 is well established as a standard, there is upwards pressure on the number of channels all of the time, at least for specialized purposes. For this reason, among others, many large-format consoles use a basic eight main bus structure. This permits a little "growing room" for the future, or simultaneous 5.1-channel and 2-channel mix bussing.

On large film and television consoles, the multichannel bus structure is available separately for dialogue, music, and effects, making large consoles have 24 main output busses. The 8-bus structure also matches the 8-track digital multitrack machines, random access hard disc recorders, and DAW structures that are today's logical step up from 2-channel stereo.

Auxiliary sends are normally used to send signals from input channels to outboard gear that process the audio. Then the signal is returned to the main busses through auxiliary returns. Aux sends can be pressed into use as output channel sends for the surround channels, and possibly even the center. Some consoles have 2-channel stereo aux sends that are suitable for left surround/right surround duty. All that is needed is to route the aux send console outputs to the correct channels of the output recorder, and to monitor the channels appropriately.

Piping multichannel digital sound around a professional facility is most often performed on AES-3 standard digital audio pairs arranged in the same order as the tape master, described below. A variant from the 110 ohm balanced system using XLR connectors that is used in audio-for-video applications is the 75 ohm unbalanced system with BNC connectors to standard AES-3id. This has advantages in video facilities as each audio pair looks like a video signal, and can be routed and switched just like video.

Even digital audio routing is subject to an analog environment in transmission. Stray magnetic fields add "jitter" at the rate of the disturbance, and digital audio receiving equipment varies in its ability to reject such jitter. Even cable routing of digital audio signals can cause such jitter; for instance, cable routed near the back of CRT video monitors is potentially affected by the magnetic deflection coils of the monitor, at the sweep rate of 15.7 kHz for standard definition NTSC video. Digital audio receivers interact with this jitter up to a worst case of losing lock on the source signal. It may seem highly peculiar to be waving a wire around the back of a monitor and have a digital audio receiver gain and lose lock, but that has happened.

Track Layout of Masters

Due to a variety of needs, there is more than one standardized method of laying out tracks with an 8-channel group on a DAW or a DTRS-style tape. One of the formats has emerged as preferred through its adoption on digital equipment, and its standardization by multiple organizations. It is given in Table 4-1.

Table 4-1 Track Layout of Masters

Track	1	2	3	4	5	6	7	8
Channel	L	R	C	LFE	LS	RS	Option	Option

Channels 7 and 8 are optionally a matrix encoded left total, right total (Lt/Rt) pair, or they may be used for such alternate content as mixes for the hearing impaired (HI) or visually impaired (VI) in television use. For 20-bit masters they may be used in a bit-splitting scheme to store the additional bits needed by the other 6 tracks to extend them to 20 bits, as described on page 100. Since there are a variety of uses of the "extra" tracks, it is essential to label them properly.

This layout is standardized within the International Telecommunications Union (ITU) and Society of Motion Picture and Television Engineers (SMPTE) for interchange of program content accompanying a picture. The Music Producer's Guild of America (MPGA) has also endorsed it.

Two of the variations that have seen more than occasional use are given in Table 4-2.

Table 4-2 Alternate Track Layout of Masters

Track	1	2	3	4	5	6	7	8
Film use	L	LS	C	RS	R	LFE	Option	Option
DTS music	L	R	LS	RS	C	LFE	Option	Option

Double-System Audio with Accompanying Video

Most professional digital videotape machines have 4 channels of 48 kHz sample rate linear pulse code modulation (LPCM) audio, and are thus not suitable for direct 5.1-channel recording. In postproduction a format based originally on 8 mm videotape, DTRS (often called DA-88 for the first machine to support the format) carrying only digital audio is often used having 8 channels capability. Special issues for such double-system recordings especially include synchronization by way of SMPTE time code. The time code on the audiotape must match that on the videotape as to frame rate (usually 29.97 fps), type (whether drop frame or non-drop frame, usually drop frame in television broadcast operations), and starting point (usually 01:00:00:00 for first frame of program).

Reference Level for Multichannel Program

Reference level for digital recordings varies in the audio world from –20 dBFS to as high as –12 dBFS. The SMPTE standard for program material accompanying video is –20 dBFS. The EBU reference level is –18 dBFS. The trade-offs among the various reference levels are:

- –20 dBFS reference level was based on the performance of magnetic film, which may have peaks of even greater than +20 dB above the analog reference level of 185 nWb/m, which is standard. So for movies transferred from analog to digital, having 20 dB of headroom was a minimum requirement, and on the loudest movies some peak limiting is necessary in the transfer from analog to digital. This occurs not only because the headroom on the media is potentially greater than 20 dB, but also because it is commonplace to produce master mixes separated into "stems," consisting of dialogue, sound effects, and music multichannel elements. The stems are then combined at the print master stage, increasing the headroom requirement.
- –12 dBFS reference level was based on the headroom available in some analog television distribution systems, and the fact that

television could use limiting to such an extent that losing 8 dB of headroom capability was not a big issue. This was based on the fact that analog television employs lots of audio compression to get programs and commercials, and station-to-station changes, to interchange better than if more headroom were available. Low headroom implies the necessity for limiting the program. Digital distribution does not suffer the same problems, and methods to overcome source-to-source differences embedded in the distribution format are described in Chapter 5.

- –18 dBFS was chosen by the EBU apparently because it is a simple bit shift from full scale. That is, –18 dB (actually –18.06 dB), bears a simple mathematical relationship to full scale when the representation is binary digits. This is one of two major issues in the transfer of movies from NTSC to PAL; an added 2 dB of limiting is necessary in order to avoid strings of full-scale coded value (hard clipping). (The other major issue is the pitch shift due to the frame rate difference. Often ignored, in fact the 4% pitch shift is readily audible to those who know the program material, and should be corrected.)

An anomaly in reference level setting is that as newer, wider dynamic range systems come on line, the reference levels have not changed; that is, all of the improvement from 16-bit to 20-bit performance, from 93 to 117 dB of dynamic range, has been taken as a noise improvement, rather than splitting the difference between adding headroom and decreasing noise.

Fitting Multichannel Audio onto Digital Video Recorders

It is very inconvenient in network operations to have double-system audio accompanying video. Since the audio carrying capacity of digital videotape machines is only 4 channels of LPCM, there is a problem. Also, since highly bit rate compressed audio is pushed to the edge of audible artifacts, with concatenation of processes likely to put problems over the edge, audio coded directly for transmission is not an attractive alternative for tapes that may see added postproduction, such as the insertion of voice overs and so forth. For these reasons, a special version of the coding system used for transmission, Dolby AC-3, called Dolby E (E for editable), is available. Produced at a compression level called mezzanine coding, this codec is intended for postproduction applications, with a number of cycles of compression–decompression possible without introducing audible artifacts, and special editing features, and so forth.

Multichannel Monitoring Electronics

Besides panning, the features that set apart multichannel consoles from multibus stereo consoles are the electronic routing and switching monitor functions for multichannel use. These include:

- Source-playback switching for multichannel work. This permits listening either to the direct output of the console, or the return from the recorder, alternately. There are a number of names for this feature, growing out of various areas. For instance, in film mixing, this function is likely to be called PEC/direct switching, dating back to switching around an optical sound camera between its output (photo-electric cell) and input. The term source/tape is also used, but is incorrect for use with a hard disc recorder. Despite the choice of terminology for any given application, the function is still the same: to monitor pre-existing recordings and compare them to the current state of mixing, so that new mixes can be inserted by means of the punch-in/punch-out process seamlessly.

  In mixing for film and television with stems, a process of maintaining separate tracks for dialogue, music, and sound effects; this switching involves many tracks, such as in the range from 18 to 24 tracks, and thus is a significant cost item in a console. This occurs since each stem (dialogue, music, or effects) needs multichannel representation (L, C, R, LS, RS, LFE). Even for the stem that seems that mono would be adequate for, dialogue, has reverberation returns in all of the channels, so needs a multichannel representation.
- Solo/mute functions for each of the channels.
- Dim function for all of the channels, about –15 dB monitor plus tally light.
- Ganged volume control. It is desirable to have this control calibrated in decibels compared to an acoustical reference level for each of the channels.
- Individual channel monitor level trims. If digital, this should have less than or equal to 0.5 dB resolution; controls with 1 dB resolution are too coarse.
- Methods for monitoring the effects of mixdown from the multichannel monitor, to 2 channels and even to mono, for checking the compatibility of mixes across a range of output conditions.

Multichannel Outboard Gear

Conventional outboard gear such as more sophisticated equalizers than the ones built into console channels may of course be used for multichannel work, perhaps in greater numbers than ever before.

These are unaffected by multichannel, except that they may be used for equalizing for the HRTFs of the surround channels.

Several types of outboard signal processing are affected by multichannel operation; these include dynamics units (compressors, expanders, limiters, etc.), and reverberators.

Processors affecting level may be applied to 1 channel at a time, or to a multiplicity of channels through linking the control functions of each of a number of devices. Here are some considerations:

- For a sound that is primarily monaural in nature, single-channel compressors or limiters are useful. Such sound includes dialogue, Foley sound effects, "hard effects" (like a door close), etc. The advantage of performing dynamics control at the individual sound layer of the mix is that the controlling electronics is less likely to confuse the desired effect with overprocessing multiple sounds. That is, if the gain control function of a compressor is supposed to be controlling the level of dialogue, and a loud sound effect comes along and turns down the level, it will turn down the level of the dialogue as well. This is undesirable since one part of the program material is affecting another. Thus, it is better to separately compress the various parts and then put them together, rather than to try to process all of the parts at once.
- For spatialized sound in multiple channels, multiple dynamics units are required, and they should be linked together for control (some units have an external control input that can be used to gang more than two units together). The multiple channels should be linked for spatialized sound because, for example, not to do so leads to one compressed channel—the loudest—being turned down more than the other channels: this leads to a peculiar effect where the subdominant channels take on more prominence than they should have. Sometimes this sounds like the amount of reverberation is "pumping," changing regularly with the signal, because the direct (loudest) to reverberant (subdominant) ratio is changing with the signal. At other times, this may be perceived at the amount of "space" changing dynamically. Thus, it is important to link the controls of the channels together.
- In any situation in which matrixed Lt/Rt sound may be derived, it is important to keep the 2 channels well matched both statically and dynamically, or else steering errors may occur. For instance, if stereo limiters are placed on the 2 channels and one is set with a lower threshold than the other accidentally, for a monaural centered sound that exceeds the threshold of the lower limiter, that sound will be allowed to go higher on the opposite channel, and

the decoder will "read" this as dominant, and pan the signal to the dominant channel. Thus, steering errors arise from mismatched dynamics units in a matrixed system.

Reverberators are devices that need to address multichannel needs, since reverberation is by its nature spatial, and should be available for all of the channels. As described above, reverberation returns on the front channels indicate listening into a space in front of us, while reverberation returns on all of the channels indicates we are listening in the space of the recording. If specific multichannel reverberators are not available, it is possible to use two or more stereo reverbs, with the returns to the 5 channels, and with the programs typically set to similar, but not identical, parameters.

Decorrelators are valuable additions to the standard devices available as outboard gear in studios, although not yet commonplace. There are various methods to decorrelate, some of them available on multipurpose digital audio reverberation devices. They include the use of a slight pitch shift (works best on non-tonal ambience), chorus effects, complementary comb filters, etc.

Inter-track Synchronization

Another requirement that is probably met by all professional audio gear, but that might not be met by all variations of computer audio cards or computer networks, is that the samples remain absolutely synchronous across the various channels. This is for two reasons. The first is that one sample at a sample rate of 48 kHz takes 20.8 μs, but one just noticeable difference psychoacoustically is 10 μs, so if 1 channel suffers a one sample shift in time, the placement of phantom images between that channel and its adjacent ones will be affected (see Chapter 6). The second is that, if the separate channels are mixed down from 5.1 to 2 channel in some subsequent process, such as in a set-top box for television, a one sample delay between channels summed at equal level will result in a notch in the frequency response of the common sound at 12 kHz, so that a sound panned from 1 channel to another will undergo a notched response when the sound is centered between the two, and will not have the notch when the pan is at the extremes—an obvious coloration.

Multichannel audio used for surround sound has a plurality of channels, yet when conventional thinking normally applied to stereo is used for the ingredient parts of a surround mix several problems emerge. Let's take as an example the left and right surround channels, designated LS and RS. Treated as a pair for the purposes of digital audio and delivered on one AES-3 cable, one could think that to produce good practice one should apply a phase correlation meter or an oscilloscope Lissajous display to show the phase relationship between the 2 channels. From a Tektronix manual:

> Phase Shift Measurements: One method for measuring phase shift—the difference in timing between two otherwise identical periodic signals—is to use XY mode. This measurement technique involves inputting one signal into the vertical system as usual and then another signal into the horizontal system—called an XY measurement because both the X and Y axis are tracing voltages. The waveform that results from this arrangement is called a Lissajous pattern (named for French physicist Jules Antoine Lissajous and pronounced LEE–sa–zhoo). From the shape of the Lissajous pattern, you can tell the phase difference between the two signals ... The measurement techniques you will use will depend on your application.[1]

Precisely. That is, one could apply the measurement technique to a recorder, say, to be certain that it is recording "in phase," and this would be good practice, but if one were to apply a Lissajous requirement for a particular program's being "in phase" then the result would not be surround sound! The reason for this is that if in-phase content is heard over two surround monitor channels that are acoustically balanced precisely, with identical loudspeakers and room acoustics, and one listens sitting exactly on the centerline and facing forward, what is heard is not surround sound at all, but inside the head sound like that produced by monaural headphones. So to apply a phase correlation criteria to surround program material is counterproductive to the whole notion of surround sound.

So a distinction has to be made between what the component parts of the system do technically, and what the phase and time relationships are among the channels of program material. The consoles, recorders, and monitor systems must maintain certain relationships among the channels to be said to be working properly, while the program material has a quite different set of requirements. Let us take up first the requirements on the equipment, and then on the program.

---

[1]www.tek.com/Measurement/App_Notes/ XYZs/measurement_techniques.pdf

Requirements for Equipment and Monitor Systems

1. All channels are to have the same polarity of signals throughout, said to be wired "in phase"; no channel may be "out of phase" with respect to any other, for well-known reasons. This applies to the entire chain. Note that AES-3 pairs actually are polarity independent and could contain a wiring error without causing problems because it is the coding of the audio on the interface, not the wiring, that controls the polarity of the signals.
2. All channels shall have the correct absolute polarity, from microphone to loudspeakers. Absolute polarity is audible, although not prominent, because human hearing contains a mechanism akin to half-wave rectification, and such a rectifier responds differently to positive-going wavefronts than to negative-going ones. For microphones this means pin 2 on its XLR connector shall produce a positive output voltage for a positive-going sound compression wave input. Caution: measurement microphones such as Bruel & Kjaer ones traditionally have used the opposite polarity, so testing systems with them must take this into account. For loudspeakers this means that a positive-going voltage shall produce a positive pressure increase in front of the loudspeaker, at least for the woofer. (Note that some loudspeaker crossover topologies force mid-range or tweeters to be wired out of phase with respect to the woofer to produce correct summing through the crossover region. Other topologies "don't care" about polarity of drivers (such types have 90° phase shifts between woofer and say mid-range at the crossover frequency). One could easily think that those topologies that result in requiring mid-ranges and tweeters to be wired in phase might be "better" than those wired out of phase. Also note that some practice is opposite to this. JBL Professional loudspeaker polarity was originally set by James B. Lansing to producing rarefaction (negative-going pressure) from the woofer when a positive voltage was applied to the loudspeaker system. In more recent times, JBL has switched polarity of their professional products to match the broader range of things on the market, and even their own professional products in other markets. For polarity of their models, see www.jblpro.com >Technical Library >Tech Note Volume 1, #12C.
3. Note that some recorders will invert absolute polarity while, for instance, monitoring their input, but have correct absolute polarity when monitoring from tape. Consoles may have similar problems for insertion paths, for instance. All equipment should be tested for correct absolute polarity by utilizing a half-wave rectified sine wave, say positive going, and observing all paths and switching conditions for maintaining positive polarity on an oscilloscope.

4. All channels shall be carried with identical mid-range latency or time delay, with zero tolerance for even single sample offsets among or across the channels. Equipment should be tested to ensure that the outputs are being delivered simultaneously from an in-phase input, among all combinations of channels. See an article on this that quotes the author extensively on the Crystal Semiconductor web site: http://www.cirrus.com/en/support/design/whitepapers.html. Download the article Green, Steven "A New Perspective on Decimation and Interpolation Filters".
5. All channels should be converted on their inputs and outputs with the same technology conversion devices (anti-alias and anti-image filters, basic conversion processes) so that group delay versus frequency across the channels is identical. (Group delay is defined as the difference in time among the various parts of the spectrum. The mid-range time of conversion is called latency, and must also be identical. Note that some converter manufacturers confuse the two and give the term group delay when what is really meant is latency.) The inter-channel phase shift is more audibly important than the monophonic group delay, since inter-channel phase shifts lead to image shifts, whereas intra-channel delay has a different mechanism for audibility. Preis has studied this extensively, with a meta paper reviewing the literature.[2] Modern day anti-aliasing and anti-imaging filters have inaudibly low group delay, even for multiple conversions, as established by Preis. Any audible group delay is probably the result of differences among the channels as described above.
6. The most common monitoring problem is to put L, C, and R monitor loudspeakers in a line and then listen from the apex of an equilateral triangle formed by left and right loudspeakers and the listening location. This condition advances the center channel in time due to its being closer, the amount of which is determined by the size of the triangle. Even the smallest amount of leading time delay in center makes pans between it and adjacent channels highly asymmetrical. What you will hear is that as soon as a pan is begun the center channel sticks out in prominence, so that the center of the stereo field is "flattened," emphasizing center. This is due to the precedence effect that the earlier arriving sound will determine the direction, unless a later occurring one is higher in level. The two

[2]Preis, multiple papers found on the www.aes.org web site under the pre-print search engine, especially "Phase Distortion and Phase Equalization in Audio Signal Processing—A Tutorial Review" AES 70th Convention, October 30–November 2, 1981. New York. Preprint 1849.

solutions to this are either to put the loudspeakers mounted on an arc with the main listening location as the center point, or to delay electrically the signal to the center loudspeaker to make it line up in time with left and right.

7. Likewise left and right surround speakers have to be at the same distance from the listening location as left, center, and right or be timed to arrive at the correct time, if the program content is to be heard the way that end users will hear it. Note that most controllers (receivers) for the home, at least the better ones, contain interchannel delay adjustment for this effect, something that the studio environment would do well to emulate.

However, note that due to psychoacoustics the side phantoms are much less good than the front and back ones, and tend to tear apart, with part of the noise heard in say each of the right and right surround channels instead of a coherent side phantom. The way we check is to rotate while listening, treating each pair as a stereo pair for these purposes, which helps ensure that everything is in phase.

With the foregoing requirements met, then program material may be monitored correctly for time and phase faults. Of course these time-based requirements apply along with many others for good sound. For instance, it is important to match the spectrum of the monitor system across the channels, and for that spectrum to match the standard in use.

Program Monitoring

Monitoring for issues such as phase flips in content is made more complicated in multichannel than in 2-channel stereo. That is first and foremost because of the plurality of channels: What should be "in phase" with what? What about channel pairings? Items panned halfway between either left and center, or right and center, only sound correct as a phantom image if the channels are in phase. But importantly if there is an in-phase component of the sound field between the left and right channels, then it will be rendered for a centered listener as a center phantom. This can lead to real trouble. The reason is that if there is any timing difference at all between this phantom, and the real center channel content, then comb filtering will occur.

Let's say that a mixer puts a vocalist into center, and also into left and right down say 6 dB, called a "shouldered" or "divergence" mix. If you shut off center, you will hear the soloist as a phantom down 3 dB (with power addition as you are assumed to be in the reverberant-field dominated region; different complications arise if you are in the direct-field dominated space). Only 3 dB down, the phantom image has a different frequency response from the actual center loudspeaker. This is because of acoustical crosstalk from the left loudspeaker first striking the left ear, then about 200 μs later reaching the right ear, with diffraction about the head and interaction with the pinnae in both cases occurring. For a centered phantom the opposite path also occurs of course. The 200 μs delay between the adjacent and opposite side loudspeakers and acoustical summing causes a notch in the frequency response around 2 kHz, and ripples in the response above this frequency. Now when added to an actual center loudspeaker signal, and only 3 dB down, and with a different response, the result is to color the center channel sound, changing its timbre.

Some mixers prefer the sound of a phantom image to that of an actual center loudspeaker. This is due to long practice in stereo. For instance, microphones are routinely chosen by recording vocalists and then panning them to the center of a 2-channel stereo monitor rig, that suffers from the 2 kHz dip. In a kind of audio Darwinism survival of the fittest, microphones with 2 kHz range presence peaks just happen to sell very well. Why? Because they are overcoming a problem in stereo. When the same mic. is evaluated over a 5.1-channel system and panned to center, it sounds peaky, because the stereo problem is no longer there.

It is a danger to have much in-phase content across all three front channels, as the inevitable result, even when the monitor system is properly aligned, is to produce noticeable and degrading timbre changes.

At Lucasfilm I solved this potential problem by designing and building a pan pot that did not allow for sound to be sent to all three front channels. Copied widely in the industry (and with credit from Neotek but not from others using the circuit), by now thousands of movies have been mixed with such a panner. Basically I made it extremely difficult to put the same sound in all three front channels simultaneously: one would have to patch to get it to happen.

The foregoing description hopefully helps in listening monitoring. Conventional phase meters and oscilloscopes have their place in testing the equipment in the system, but can do little today to judge program content, as there are so many channels involved, and whereas we've seen that conventional thinking like "left and right should be in phase" can cause trouble when applied to left and right fronts, or to left and right surrounds.

Postproduction Formats

Before the delivery formats in the production chain, there are several recording formats to carry multichannel sound, with and without accompanying picture. These include standard analog and digital multitrack audio recorders, hard disc-based workstations and recorders, and video tape recorders with accessory Dolby E format adapters for compressing 5.1-channel sound into the space available on the digital audio channels of the various videotape format machines.

Track Layout

Any professional multitrack recorder or DAW can be used for multichannel work, so long as other requirements such as having adequate word length and sample rate for the final release format as discussed above, and time code synchronization for work with an accompanying picture, are respected. For instance, a 24-track digital recorder could be used to store multiple versions of a 5.1-channel mix as the final product from an elaborate postproduction mix for a DVD. It is good practice at such a stage to represent the channels according to the ultimate layout of channel assignments, so that the pairing of channels that takes place on the AES-3 interconnection interface is performed according to the final format, and so that the AES pairs appear in the correct order. For this reason, the preferred order of channels for most purposes is L, R, C, LFE, LS, RS. This order may repeat for various principal languages, and there may also be Lt/Rt stereo pairs, or monaural recordings for eventual distribution of HI and VI channels for use with accompanying video. So there are many potential variations in the channel assignments employed on 24-, 32-, and 48-track masters, but the information above can help to set some rules for channel assignments.

If program content with common roots is ever going to be summed, then that content must be kept synchronized to sample accuracy. The difficulty is that SMPTE time code only provides synchronization to within 20 samples, which will cause large problems downstream if, for example, an HI dialogue channel is mixed with a main program, also containing the dialogue, for greater intelligibility. Essentially no time offset can be tolerated between the HI dialogue and the main mix, so they must be on the same piece of tape and synchronized to the sample, or use one of the digital 8-track machines that has additional synchronization capability beyond time code to keep sample accuracy. Alternatively be certain that the DAW that you use maintains sample accurate time resolution among its channels. Problems can arise when one path undergoes a different process than another. Say one path comes out of a DAW to an external device, and then back in, and the device is analog. The conversion of D to A and A to D will impose a delay that is not in the alternate paths, so if the content of this path is consequently summed with the delayed one, comb filtering will result. For instance, should you send an HI channel to an outboard compressor, and then back into your console, just the conversion latency will be enough to put it out of time with respect to the main dialogue, so that when it is combined in the user's set, comb filtering will result.

Postproduction Delivery Formats

For delivery from a production house to a mastering one for the audio-only part of a Digital Versatile Disc Video production, delivery is usually in 8 channels chunks. In the early days of this format for multichannel audio work, at least five different assignments of the channels to the tracks were in use, but one has emerged as the most widely used for sound accompanying picture. It is shown in Table 4-1 on page 122.

This layout has been standardized in the SMPTE and ITU-R. For use on most digital videotape machines, that are today limited to 4 LPCM channels sampled at 48 kHz, a special low-bit-rate compression scheme called Dolby E is available. Dolby E supplies "mezzanine" compression, that is, an intermediate amount of compression that can stand multiple cycles of compression–decompression in a postproduction chain without producing obvious audible artifacts. Using full Dolby Digital compression at 384 kbits/s for 5.1 channels runs the risk of audible problems should cascading of encode–decode cycles take place. That is because Dolby Digital has already been pressed close to perceptual limits, for the best performance over the limited capacity of the broadcast or packaged media channel. The 2 channels of LPCM on the VTRs supply a data rate of 1.5 Mbps, and therefore much less bit-rate reduction is needed to fit 5.1 LPCM channels into the 2-channel space on videotape than into the broadcast or packaged media channel. In fact, Dolby E provides up to 8 coded channels in one pair of AES channels of videotape machines. The "extra" 2 channels are used for Lt/Rt pairs, or for ancillary audio such as channels for the HI or VI. Another feature that distinguishes Dolby E from Dolby Digital broadcast coders is that the frame boundaries have been rationalized between audio and video by padding the audio out so it is the same length as a video frame, so that a digital audio-follow-video switcher can be used and not cause obvious glitches in the audio. A short crossfade is performed at an edit, preventing pops, and leading to the name Dolby "Editable." Videotape machines for use with Dolby E must not change the bits from input to output, such as sample rate converting for the difference between 59.94 and 60 Hz video.

In addition to track layout, other items must be standardized for required interchangeability of program material, and to supply information about metadata from postproduction to mastering. One of the items is so important that it is one of the very few requirements which the

FCC exercises on digital television sets: they must recognize and control gain to make use of one of the three level setting mechanisms, called dialogue normalization (dialnorm). First, the various items of metadata are described in Chapter 5, then their application to various media.

Surround Mixing Experience

With all the foregoing in mind, here are some tips based on the experience of other surround mixers and myself. The various recommendations apply more or less to various types of surround sound mixing, like the direct/ambient approach and the sound-all-round approach and to various number of channels. Where differences occur they will be noted.

Mixing all of the channels for both direct sound and reverberation at one and the same time is difficult. It is useful to form groups, with all the direct sound of instruments, main and spot microphones, in one group; and ambience/reverberation-oriented microphones and all the returns of reverberation devices on another. These groups are both for the solo function and for the fader function as we shall see. Of course if the program is multilayered especially into stems like dialogue, music, and effects, then each of these may need the same treatment. On live mixes too it is useful to have solo monitor groups so that internal balances within a given type of effect, like audience reaction, can be performed without interference from the main voice over.

First pan the source channels into their locations if they are to be fixed in space. There is little point in equalizing before panning since location affects timbre. Start mixing by setting an appropriate level and balance for the main microphone system, if the type of mix you are doing has one. For a pan pot stereo mix, it is typical to start with the main voice as everything else will normally be referenced off the level of this source. For multichannel arrays, spaced omnis and the Fukada array will use similar level across all three main mikes, typically outrigger mikes in spaced microphone stereo will be set around –5 dB relative to main microphones, and other arrays are adjusted for a combination of imaging across the front and adequate spread.

If the main array is to be supplemented by spot mikes, before setting a balance time their arrival. The best range is usually 20–30 ms after the direct sound, but this depends on the style of music, any perception of "double hits" or comb filters, and so forth. It is easy to forget this step, and I find the sound often to be muddy and undefined until these time delays are put in. Some consoles offer delay, but not this much. Digital audio editing workstations can grab whole tracks and shift them, and this may be done, or inserted delays can be employed. In conventional mixing on an analog console without adjustable delays available, the problem is that as one raises the level of a spot mike, two things are happening at once: the sound of the spot mike instrument is arriving earlier, and its level is being increased. Thus you will find the level to be very critical, and you will probably feel that any level you set is a compromise, with variation in apparent isolation of the instrument with its level. This problem is ameliorated with correct timing.

However, at least one fine mixer finds that the above advice on timing is not necessary with spaced omni recordings (which tend to be spacious but not very well imaged). With this type of main mike, the extra imaging delivered by the earlier arrival of the spot mike can help.

The spot mikes will probably be lower in level than the main microphones, just enough the "read" the instrument more clearly. During production of a London Decca recording of the Chicago Symphony in the Great Hall of the Krannert Center in Urbana, Illinois that I observed many years ago, the producer played all of the available competitive records and made certain that internal orchestral balances that obscured details in the competitive recordings would be heard in the new one. It means that there may be some riding of gain on spot mikes, even with delays, but probably less than there would have been without the availability of delay.

With the main and spot microphones soloed, get a main mix. Presumably at this stage it will, and probably should, sound too dry, lacking in the warmth that reverberation adds. Activate aux sends for reverberation as needed to internal software or external devices. Main microphone channel pairs, such as an ORTF pair, are usually routed by stereo aux busses to stereo reverberation device inputs. If the reverberator only has 2 channels of output, parallel the inputs of two such devices and use the returns of the first for L/R and the second for LS/RS. Set the two not to identical but to nearly identical settings so that there is no possibility of phantom images being formed by the outputs of the reverberators.

In film mixing, the aux sends are separated by stem: dialogue, music, and effects are kept separate. This is so later on M&E mixes can be pulled from the master mix for foreign language dubs. Route the output of the reverberation devices, or the reverberant microphone tracks to the multichannel busses, typically L/C/R/LS/RS and potentially more.

Now solo all the ambient/reverberation microphones and/or reverberator outputs. Be certain that the aux sends of the main and spot mikes are not muted by the solo process. Build a reverberant space so that it sounds enveloping, spacious, and without particular direction. Then for direct/ambient recording, bias the total reverberant field by something like 2–3 dB to the front (this is because the frontal sources will tend to mask the reverberation more than from other directions). The reverberant field at this point will sound "front heavy," but that is probably as it should be. For source-all-round approaches to mixing, this consideration may not apply, and decorrelated reverberation should probably appear at equal level in L/R/LS/RS. If more channels are available by all means use them. Ando has found (see Chapter 6) that five is the minimum number of channels to produce a diffuse sound field like reverberation, however, the angles for this were ±36°, ±108°, and +180° from straight ahead. While ±36° can be approximated with left and right at ±30°, and ±108° easily by surrounds in the ±100–120° range of the standard, the center back channel is not available in standard 5.1. However, it is available in Dolby's Surround EX and DTS's Surround ES, so is the next channel to be added to 5.1.

Now using fader groups, balance the main/front mikes against the reverberant-field sources. Using fader groups locks the internal balances among the mike channels in each category, and thus makes it easy to maintain the internal balances consistently, while adding pleasant and balanced reverberation to the mix. You will probably find at this stage that the surround level is remarkably sensitive, with ±1 dB variation going from sounding all up front to sounding surround heavy. Do not fear, this is a common finding.

If an Lt/Rt mixdown is to be the main output, be certain to monitor through a surround encoder/decoder pair as the width of the stereo stage will interact with the microphone technique. Too little correlation and the result will be decoded as surround content; too much and mono center will be the result.

One Case Study: Herbie Hancock's "Butterfly" in 10.2

It has already been explained that this piece was written with surround sound in mind, but when delivered as a 2-track CD mix due to the record company's demand it failed in the marketplace, probably because too much sound was crammed into too little space. Herbie Hancock has been a supporter of surround sound for many years, including writing for the medium and lending his name to the International Alliance for Multichannel Music among other things. This follows his mantra of "don't be afraid to try things." He lent us the 48-track original session files of this tune for a demonstration of 10.2-channel sound at the Consumer Electronics Show. His mixer Dave Hampton was supposed to do the work accompanied by me, but he became ill and was not available to do the work so I took it on. Ably assisted by then undergraduate Andrew Turner, we spent two very very long days producing a mix. Having done it, Herbie came down to USC and listened to it and said one of the best things I've ever been told: "Now the engineer becomes one of the musicians." So here is what I did.

First fully one-half of the mixing time was really spent in an editing function, sorting out tracks, muting parts we didn't want to use, for crosstalk or other reasons. This is really an editorial function, not a mixing one, but nonetheless had to be done in the mixing period. In film work, this would have been done off line, in an edit room, then brought to a mix stage for balancing.

The direct/ambient approach was inappropriate for this mix as one wants to spread the sound out the most and the sound-all-round approach offered us the most amount of articulation in the mix, the opposite of why the 2-channel version failed. Here is how the various parts of the mix were treated:

- Herbie Hancock is a keyboard player, in this case of electronic keyboards. They were set to be a rather warm sound, with blurred attacks, not like a traditional piano sound. So we decided to put them in left and right wide speakers at ±60° and left and right direct radiating surround at ±110°, so that the listener is embedded in the keyboard parts.
- This is a jazz piece, with the idiom being that each solo takes the spotlight. For us, this mean front and center, so most solos were panned to center.
- The primary solo, a flute part, was put in center front, but also in center back, just as an experiment—a play on front/back confusion.
- Percussion was largely kept in front LCR because percussion in the surround we find distracting to the purpose: it spotlights the speaker positions.
- Certain effects sounding parts I would call zings were put in center back, to highlight them.
- Hand chimes were put in left and right height channels, at ±45° in plan and 45° elevated, and panned between the 2 channels as glissandos were played on them.
- At the end, when the orchestration thins out to be just the flute solo, the flute "takes off" and flys around the room.

One primary thought here is that while it is possible to pan everything all the time, too many things cannot be panned, as that would result in confusion and possibly dizziness. Keeping the amount of panning smaller keeps it perceptible in a good way. And with *Butterfly*, as it turned out, there was a good reason to pan the flute at the end: Herbie told us that it was the butterfly and it takes off at the end, something that frankly had escaped us in getting buried in mixing!

Surround Mixing for DVD Music Videos

Music videos must compete with 2-channel mixes, even though they are in surround. This is due to the comparison of the surround mix with the existing 2-channel one during postproduction of music videos. With dialnorm (see p. 154) adjusting the level to make the source more interchangeable with other sources downwards by something on the order of 7 dB, the mix seems soft to producers and musicians. The comparison of a 2-channel mix recorded with flattened peak levels through limiting with a wider dynamic range mix lowered by the amount necessary so that so much limiting is not necessary and by the amount necessary to make it comparable to other sources in the system (applying dialnorm), is unfavorable to the surround mix. Thus it may wind up being even more compressed/limited than the 2-track mix, or at least as much. The bad practices that have crept into music mastering over the years are carried across to media with very wide dynamic range capacity, only a tiny fraction of which is used.

Another factor in music mixing is the use of the center channel. Since many mix engineers have a great deal of experience with stereo, and are used to its defects and have ways around them, several factors come into play. The most far out don't use the center at all. Those that do tentatively stick one toe in the water and might put the bass fundamentals there (they've been taught that to prevent "lifts" in LP production: areas where the cutting stylus might retract so far that it doesn't produce a groove—obviously of no relevance to digital media). Or only the lead solo might be put in the center, leading to paranoid reactions of the artist when they find someone on the other end can solo their performance. In the best film mixing, all three front channels are treated equally for all the elements. Music is recorded in 3-channel format for film mixes. B movies may well use needle-drop music off CD in left and right only in the interest of time and money saving, but it is bad practice that should be excoriated.

An example is in order. We built a system for 2-channel stereo reproduction that separately processed the center channel (from the center of a 5.1 input, or from a 2-channel Lt/Rt input decoded into LCRS then put back together into L/R and S). The purpose of having the center channel separate was to do special signal processing on it before re-insertion as a phantom image. A television camera was arranged on top of the corresponding picture monitor (using a direct view monitor with no room for a center speaker was the original impetus of this work), looking out at the listener. Through sophisticated face recognition, the location of the person's ears could be found. By correcting the time delay of the center separately into left and right speakers, a centered sound could be kept centered despite the person leaning left and right. Naive listeners when this was explained to them had no problem with the concept, saw the point of it, and found it to work well. Professional listeners, on the other hand, were flummoxed, convinced there was some kind of black magic at work: they were so used to the defect of the phantom center moving around as one moves one's head that the sensation was uncanny to them! This example demonstrates that change only comes slowly since workarounds have been found for problems to the extent that they have become standard practice.

George Massenburg

Multi-Grammy Winner, Music Producer & Engineer, and Equipment and Studio Design Engineer

An interview with George Massenburg on surround sound and allied topics is available at www.tmhlabs.com/pub.

5 Delivery Formats

Tips from This Chapter

- The multichannel digital audio consumer media today are Digital Versatile Disc Video (DVD-V), Blu ray, HD DVD, terrestrial over-the-air and satellite broadcasting, and possible delivery of these by cable, either copper or fibre optic. Internet downloadable movies are beginning, with the requirement that consumers expect the facilities of at least one stream of audio that the competition offers.
- Metadata (data about the audio "payload" data), wrappers (the area in a digital bitstream to record the metadata), and data essence (the audio payload or program) are defined.
- Linear PCM (LPCM) has been well studied and characterized, and the factors characterizing it include sample rate (see Appendix 1), word length (see Appendix 2), and the number of audio channels. Redundancy in audio may be exploited to do bit packing much like Zip files do for documents; the underlying audio coding is completely preserved through such processes.
- Word length needs to be longer in the professional domain than on the release media, so that the release may achieve the dynamic range implied by its word length, considering the effects of adding channels together in multitrack mixing.
- Products may advertise longer word lengths than are sensible given their actual dynamic range, because many of the least significant bits may contain only noise. Table 5-1 gives dynamic range versus the effective number of bits.
- Coders other than LPCM have application in many areas where LPCM even with bit-reduction packing is too inefficient. There are several classes of such coders, with different characteristics featuring various tradeoffs of factors such as maximum bit rate reduction, ability to edit, and ability to cascade.

Table 5-1 Number of Bits versus Dynamic Range

Effective number of bits	*Dynamic range, dB**
16	93
17	99
18	105
19	111
20	117
21	123
22	129
23	135
24	141

*Includes the effect of triangular probability density amplitude function dither, preventing quantization distortion and noise modulation; this dither adds 3 dB to the noise floor to prevent such problems.

- One class of such coders, called perceptual coders, utilize the masking characteristics of human listeners in the frequency and time domains including the fact that louder sounds tend to obscure softer ones to make more efficient use of limited channel capacity. Perceptual coders tend to offer the maximum bit rate reduction.
- Multiple tracks containing content intended to make a stereo image must be kept synchronized to the sample. Even a one-sample shift is audible as a move in a phantom image between two adjacent channels.
- Reference level on professional masters varies from −20 dBFS (Society of Motion Picture and Television Engineers, SMPTE), through −18 dBFS (EBU), up to as much as −12 dBFS (some music uses).
- Many track layouts exist, but one of the most common is the one standardized by ITU (International Telecommunications Union) and SMPTE for interchange of program accompanying pictures at least. It is L, R, C, LFE (Low Frequency Enhancement), LS, RS, and 7 and 8 used variably for such ancillary uses as Lt/Rt, or Hearing Impaired (HI) and Visually Impaired (VI) mono mixes.
- Most digital video tape machines have only four audio tracks, thus need compression schemes such as Dolby E to carry 5.1-channel content (in one audio pair).
- DTV, DVD-V, HD DVD, and Blu-ray have the capability for multiple audio streams accompanying picture, which are intended to be selected by the end user.
- Metadata transmits information such as the number of channels and how they are utilized, and information about level, compression, mixdown of multichannel to stereo, and similar features.

- There are three metadata mechanisms that affect level. Dialogue normalization (dialnorm) acts to make programs more interchangeable with each other, and is required of every ATSC TV receiver. Dynamic Range Control (DRC) serves as a compression system that in selected sets may be adjusted by the end user. Mixlevel provides a means for absolute level calibration of the system, traceable to the original mix. When implemented all three tend to improve on the conditions of NTSC broadcast audio.
- There is a flag to tell receiving equipment about the monitor system in use, whether X curve film monitoring, or "flat" studio and home monitoring. End-user equipment may make use of this flag to set playback parameters to match the program material.
- The 2-channel mode can flag the fact that the resulting mix is an Lt/Rt one intended for subsequent matrix decoding, or is conventional stereo, called Lo/Ro.
- Downmix sets parameters for the level of center and surrounds to appear in Left/Right outputs.
- Film mixes employ a different standard from home video. Thus, transfers to video must adjust the surround level down by 3 dB.
- Sync problems between sound and picture are examined for DVD-V and DTV systems. There are multiple sources of error that can even include the model of player.
- Each of the features of multichannel digital audio described above has some variations when applied to DVD-V and Digital Television.
- Intellectual property protection schemes include making digital copies only under specified conditions and watermarking so that even analog copies derived from digital originals can be traced.

Introduction

There are various delivery formats for multichannel audio available today, for broadcast media, packaged media, and downloadable media. Most of the delivery formats carry, in addition to the audio, metadata, or data about the audio data, in order that the final end-user equipment be able to best reproduce the producer's intent. So in the multichannel world, not only does information about the basic audio, such as track formats, have to be transmitted from production or postproduction to mastering and/or encoding stages, but also, information about how the production was done needs to be transmitted. While today such information has to be supplied in writing so that the person doing the mastering or encoding to the final release format can "fill in the blanks" about metadata at the input of the encoder, it is expected that this transmission of information will take place electronically in the future, as a part of a database accompanying the program. Forms that can be used to transmit the information today are given at the end of this chapter.

The various media for multichannel audio delivery to consumer end users, and their corresponding audio coding methods, in order of introduction, are as follows:

- *Laser Disc*: Dolby Digital, Digital Theater Systems (DTS).
- *DTS CD*: DTS Digital Surround (see Appendix 3).
- *US Digital Television*: Dolby Digital.
- *Digital Versatile Disc Video (DVD-V)*: LPCM (Linear PCM), Dolby Digital, DTS, MPEG-2 Musicam Surround.
- *Super Audio Compact Disc (SACD)*: DSD.
- *Digital Versatile Disc Audio (DVD-A)*: LPCM, LPCM with MLP bit packing, and others (see Appendix 3).
- *HD DVD*: LPCM, Dolby Digital, Dolby Digital Plus, Dolby TrueHD, DTS Digital Surround, DTS-HD High Resolution Audio, DTS-HD Master Audio.
- *Blu Ray*: LPCM, Dolby Digital, Dolby Digital Plus, Dolby TrueHD, DTS Digital Surround, DTS-HD High Resolution Audio, DTS-HD Master Audio.
- *Internet Downloadable Video and Audio*: Various codecs.
- *Digital Cinema*: LPCM.
- There are proposals for multichannel digital radio. Some of them utilize subcarriers on existing radio stations and rely on Lt/Rt matrix encoding to offer a soft failure by reversion to analog when reception conditions warrant.

This chapter begins with information about new terminology, audio coding, sample rate, and word length requirements in postproduction (supplemented with Appendixes 1 and 2, respectively), inter-track synchronization requirements, and reference level issues which the casual reader may wish to skip.

New Terminology

In today's thinking, there is a lot of new terminology in use by standards committees that has not yet made it into common usage. The audio payload of a system, stripped of metadata, error codes, etc., is called data essence in this new world. I, for one, don't think practitioners are ever going to call their product data essence, but that is what the standards organizations call it. Metadata, or data about the essence, is supplied in the form of wrappers, the place where the metadata is stored in a bit stream.

Audio Coding

Audio coding methods start with LPCM, the oldest and most researched digital conversion method. The stages in converting an analog audio signal to LPCM include anti-alias filtering,[1] then sampling the signal in the time domain at a uniform sample rate, and finally quantizing or "binning" the signal in the level domain to the required word length. A uniform step-size device called a quantizer assigns the signal a number for the closest step available with the addition of dither to linearize the "steps" of the quantizer. Appendix 1 discusses sample rate and anti-aliasing; Appendix 2 explains word length and quantization. LPCM is usually preferred for professional use up to the point of encoding for the release channel because, although it is not very efficient, it is the most mathematically simple for such processes as equalization, compared to other digital coding schemes. Also, the consequences of performing processes such as mixing are well understood so that requirements can be set in a straightforward manner. For instance, adding two equal, high-level signals together can create a level that is greater than the coding range of one source channel, but the amount is easily predicted and can be accounted for in design by "turning the signals down" before addition. Likewise, the addition of noise due to adding source channels together is well understood, along with requirements for re-dithering level-reduced signals to prevent quantization distortion of the output DAC (see Appendix 2), although all equipment may not be well designed with such issues in mind.

One alternate approach to LPCM is called 1-bit $\Delta\Sigma$ (delta-sigma) conversion. Such systems sample the audio at a much higher frequency than conventional systems, such as at 2.82 Mbits/s (hereafter Mbps) with 1-bit resolution to produce a different set of conversion tradeoffs than LPCM. Super Audio CD by Sony and Philips use such a system. SD systems generally increase the bit rate compared to LPCM, so are of perhaps limited appeal for multichannel audio.

LPCM is conceptually simple, and its manipulation is well understood. On the other hand, it cannot be said to be perceptually efficient, because only its limits on frequency and dynamic ranges are adjusted to human hearing, not the basic method. In this way, it can be seen that even LPCM is "perceptually coded," that is, by selecting a sample rate,

[1]A steep filter is required so that content at greater than 1/2 the sample rate is excluded from the conversion process. For example, if an input 28-kHz tone were not suppressed in a system employing 48-kHz sampling, that tone would be rendered as a 20 kHz one on the output! Effectively higher frequencies than one-half the sample rate "bounce" off the folding frequency, which is 1/2 the sample rate, and wind up at lower frequencies.

word length, and number of channels, one is tuning the choices made to human perception. In fact, DVD-A gives the producer a track-by-track decision to make on these three items, factors that can even be adjusted on the basis of particular channel combinations.

A major problem for LPCM is that it can be said to be very inefficient when it comes to coding for human listeners. Since only the bounds are set psychoacoustically, the internal coding within the bounds could be made more efficient. More efficient coding could offer the representation of greater bandwidth, more dynamic range, and/or a larger number of channels within the constraints of a given bit rate. Thus, it may well be that the "best" digital audio method for a given channel is found to be another type of coding; for now LPCM is a conservative approach to take in original recording and mixing of the multigenerations needed in large-scale audio production. This is because LPCM can be cascaded for multiple generations with only known difficulties cropping up, such as the accumulation of noise and headroom limitations as channels are added together. In the past, working with multigeneration analog could sometimes surprise listeners with how distorted a final generation might sound compared to earlier generations.[2] What was happening was that distortion was accumulating more or less evenly among multiple generations, so long as they were well engineered, but at one particular generation the amount of distortion was "going over the top" and becoming audible. There is no corresponding mechanism in multiple generation LPCM, so noise and distortion do not accumulate except for the reasons given above.

A variety of means have been found to reduce the bit rate delivered by a given original rate of PCM conversion to save storage space on media or improve transmission capability of a limited bit rate channel. It may be viewed that within a given bit rate, the maximum audio quality may be achieved with more advanced coding than LPCM. These various means exploit the fact that audio signals are not completely random, that is, they contain redundancy. Their predictability leads to ways to reduce the bit rate of the signal. There is a range of bit-rate-reduction ratios possible, called "coding gain," from as little as 2:1 to as much as 15:1,

[2]It did surprise me. The 70 mm prints of *Return of the Jedi* had audible IM distortion, and yet an improved oxide was in use compared to earlier prints. I worked hard to understand why, when this generation measured about the same as the foregoing analog generations, that it sounded distorted. What I concluded was that each generation was adding distortion and yet remained inaudible until a threshold was crossed in this final generation to audibility. One year later with a much improved oxide for the postproduction magnetic generations on *Indiana Jones and the Temple of Doom* the audible IM distortion was gone from the 70 mm prints, even though it contained a passage of chorus over strong bass that would normally reveal IM distortion well.

and a variety of coders are used depending on the needs of the channel and program. For small amounts of coding gain completely reversible processes are available, called lossless coders, that use software programs like computer compression/decompression systems such as ZIP to reduce file size or bit rate.

One method of doing lossless compression is Meridian Lossless Packing, MLP. It provides a variable amount of compression depending on program content, producing a variable bit rate, in order to be the most efficient on media. The electrical interfaces to and from the media, on the other hand, are at a constant bit rate, to simplify the interfaces. Auxiliary features of MLP include added layers of error coding so that the signal is better protected against media or transmission errors than previous formats, support for up to 64 channels (on media other than DVD-A that has its own limits), flags for speaker feed identification, and many others.

In order to provide more channels at longer word lengths on a given medium such as the Red Book CD with its 1.411-Mbps payload bit rate, redundancy removal may be accomplished with a higher coding gain than lossless coders using a method such as splitting the signal up into frequency bands, and using prediction from one sample to the next within each band. Since sample-by-sample digital audio tends to be highly correlated, coding the difference between adjacent samples with a knowledge of their history instead of coding their absolute value leads to a coding gain (the difference signals are smaller than the original samples and thus need less range in the quantizer, i.e., fewer bits). As an example, DTS Coherent Acoustics uses 32 frequency subbands and differential PCM within each band to deliver 5.1 channels of up to 24-bit words on the Red Book CD medium formerly limited to 2 channels of 16-bit words.

A 5.1-channel, 16-bit, 48-kHz sample rate program recorded in LPCM requires 3.84 Mbps of storage, and the same transfer rate to and from storage. Since the total payload capacity of a Digital Television station to the ATSC standard is 19 Mbps, LPCM would require about 20% of the channel capacity: too high to be practical. Thus bit-rate-reduction methods with high coding gain were essential if multichannel audio was to accompany video. One of the basic schemes is to send, instead of LPCM data that is the value for each sample in time, a successive series of spectrum analyses, which are then converted back into level versus time by the decoder. These methods are based on the fact that the time domain and the frequency domain are different representations of the same thing, and transforms between them can be used as the basis for coding gains; data reduction can be achieved in either domain. As the need for coding gain increases under conditions of lower available bit rates, bit-rate-reduction systems exploit the characteristics of human perceptual masking.

Human listening includes masking effects: loud sounds cover up soft ones, especially nearby the loud sound in frequency. Making use of masking means having to code only the loudest sound in a frequency range at any one time, because that sound will cover up softer ones nearby. It also means that fewer bits are needed to code a high-level frequency component, because quantizing noise in the frequency range of the component is masked. Complications include the fact that the successive spectra are taken in frames of time, and the frame time length is switched to longer for more steady state sound (with greater frequency resolution) or shorter for a more transient approach (with better time resolution). Temporal masking, sometimes called non-simultaneous masking, is also a feature of human hearing that is exploited in perceptual coders. A loud sound will cover up a soft one that occurs after it, but surprisingly also before it! This is called backwards masking, and it occurs because the brain registers the loud sound more quickly than the preceding soft one, covering it up. The time frame of backwards masking is short, but there is time enough to overcome the "framing" problem of transform coders by mostly hiding transient quantization noise underneath backwards masking.

Transform codecs like Dolby Digital, and MPEG AAC, use the effects of frequency and temporal masking to produce up to a 15:1 reduction in bits, with little impact on most sounds. However, there are certain pathological cases that cause the coders to deviate from transparency, including solo harpsichord music due to its simultaneous transient and tonal nature, pitch pipe because it reveals problems in the filter bank used in the conversion to the frequency domain, and others. Thus low-bit-rate coders are used where they are needed for transmission or storage of multichannel audio within the constraints of accompanying a picture in a limited capacity channel, downloading audio from web sites, or as accompanying data streams for backwards compatibility such as in the case of DVD-A discs for playback on DVD-V players.

Cascading Coders

When the needs of a particular channel require codecs be put in series, called concatenation or cascading, the likelihood of audible artifacts increases, especially as the coding gain increases. The most conservative approach calls for LPCM in all but the final release medium, and this approach can be taken by professional audio studios. On the other hand, this approach is inefficient, and networked broadcast operations may, for instance, utilize three types of coders: contribution, distribution, and emission. Contribution coders send content from the field to network operations; distribution from the network center to affiliates; and emission coders are only used for the final transmission to the end user. By defining each of these for the number of generations permissible and other factors, good transparency is achieved even under less than LPCM bit rate conditions. An example of a distribution coder is Dolby E, covered later in this chapter. An alternate term for distribution coder is "mezzanine coder." Such codecs may be designed for good transparency despite some number of cycles of coding and decoding.

Sample Rate and Word Length

The recorder or workstation must be able to work in the format of the final release with regard to sample rate and minimum word length. That is, it would be pointless to work in postproduction at one sample rate, and then up-convert to another for release (see Appendix 1 on sample rate). In the case of word length, the word length of the recorder or workstation has as an absolute minimum the same word length as the release. There are two reasons for this. The first is that the output dynamic range prescribed by the word length is a "window" into which the source dynamic range must fit. In a system that uses 20-bit A/D conversion, and 20-bit D/A conversion, the input and output dynamic ranges match only if the gain is unity between A/D and D/A. If the level is increased in the digital domain, the input A/D noise swamps the digital representation and dominates, while if the level is decreased in the digital domain, the output DAC may become under dithered, and quantization distortion products could increase. Either one results in a decrease to the actual effective number of bits. Equalization too can be considered to be a gain increase or decrease leading to the same result, albeit applying only to one frequency range. Also in the multiple stages of postproduction, it is expected that channels will be summed together. Summing reduces resolution, because the noise of each channel adds to the total. Also, peak levels in several source channels simultaneously add up to more than the capacity of one output channel, and either the level of the source channels must be reduced, or limiting must be employed to get the sum to fit within the audibly undistorted dynamic range of the output. Assuming unity gain summing of multiple channels (as would be done in a master mix for a film for instance, from many prepared pre-mixes), each doubling of the number of source channels that contributes to one output channel loses one-half bit of resolution (noise is, or should be, random and uncorrelated among the source channels, and at equal level two sources will add by 3 dB, 4 by 6 dB, 8 by 9 dB, and 16 by 12 dB). Thus, if a 96-input console is used to produce a 5.1-channel mix, each output channel could see contributions from

96/6 = 16 source channels, and the sum loses two bits of dynamic range (12 dB) compared to that of one source channel. With 16-bit sources, the dynamic range is about 93 dB for each source, but only 81 dB for the mixed result. If the replay level produces 105 dB maximum Sound Pressure Level per channel (typical for film mixes), then the noise floor will be 22 dB SPL, and audible. (Note that the commonly quoted 96 dB dynamic range for 16-bit audio is a theoretical number without the dither that is essential for eliminating a problem built into digital audio of quantizing distortion, wherein low-level sound becomes "buzzy"; adding proper dither without noise modulation effects adds noise to the channel, but also linearizes the quantizing process so that tones can be heard up to 15 dB below the noise floor, which otherwise would have disappeared.) Thus, most digital audio workstations (DAWs) employ longer word lengths internally than they present to the outside world, and multitrack recorders containing multiple source channels meant to be summed into a multichannel presentation should use longer word lengths than the end product, so that these problems are ameliorated.

Due to the summation of channels used in modern-day music, film, and television mixing, greater word length is needed in the source channels, so that the output product will be to a high standard. Genuine 20-bit performance of conversion in the ADCs and DACs employed, if it were routinely available, and high-accuracy internal representation and algorithms used in the digital domain, yields 114 dB dynamic range (with dither and both conversions included). This kind of range permits summing channels with little impact on the audible noise floor. For 0 dB SPL noise floor, and for film level playback at 105 dB maximum SPL, 114 – 105 dB = 9 dB of "summation noise" that is permissible. With 20-bit performance, 8 source channels could be added without the noise becoming audible for most listeners most of the time even in very quiet spaces. In other words, each of the 8 source channels has to have an equivalent noise level that is 9 dB below 0 dB SPL in order that its sum has inaudible noise. (The most sensitive region of hearing for the most sensitive listeners actually is about 5 dB below 0 dB SPL though.)

Note that "24-bit" converters on the market that produce or accept 24 bits come nowhere near producing the implied 141 dB dynamic range. In fact, the best converters today have 120 dB dynamic range, that is 20-bit performance. And this is the specification for a typical part, not a maximum noise floor for that part. The correct measure is the effective number of bits, which is based on the dynamic range of the converter, but not often stated. I have measured equipment with "24-bit" ADC and DAC converters that had a dynamic range of 95 dB, 16 bit performance. So look beyond the number of bits to the actual dynamic range. Table 5-1 shows the dynamic range that should be deliverable for a given number of bits, but see a more comprehensive discussion in Appendix 2 on word length.

Metadata

Metadata for broadcast media was standardized through the Advanced Television Systems Committee process. Subsequently, some of the packaged media coding systems followed the requirements set forth for broadcasting so that one common set of standards could be used to exercise control features for both broadcasting and packaged media. First, the use of metadata for broadcast, and for packaged media using Dolby Digital are described, then the details of the items constituting metadata are given.

The items of metadata used in ATSC Digital Television include the following:

- *Audio service configuration*: Main or Second program. These typically represent different primary languages. Extensive language identification is possible in the "multiplex" layer, where the audio is combined with video and metadata to make a complete transmission. This differs on DVD; see its description below.
- *Bit stream mode*: This identifies one stream from among potentially several as to its purpose. Among them is Complete Main (CM), a mix of dialogue, music, and effects. Other bit stream modes are described below.
- *Audio coding mode*: This is the number of loudspeaker channels, with the designation "number of front channels/number of surround channels." Permitted in ATSC are 1/0, 2/0, 2/1, 3/0, 3/1, 2/2, and 3/2. In addition, any of the modes may employ an optional Low Frequency Enhancement (LFE) channel with a corresponding flag, although decoders are currently designed to play LFE only when there are more than 2 channels present. The audio coding modes most likely to see use, along with typical usage, are: 1/0, local news; 2/0, legacy stereo recordings, and by means of a flag to switch surround decoders, Lt/Rt; and 3/2.
- *Bit stream information*: This includes center downmix level options, surround downmix level options, Dolby Surround mode switch, Dialogue Normalization (dialnorm), Dynamic Range Control (DRC), and Audio Production Information Exists flag that references Mixing Level and Room Type.

Audio on DVD-V differs from Digital Television in the following ways:

- On DVD-V, there are from 1 to 8 audio streams possible. These follow the coding schemes shown in Table 5-2.

Table 5-2 Service Types

Code	*Service type*
0	Main audio service: Complete Main (CM)
1	Main audio service: music and effects (ME)
2	Associated service: visually impaired (VI)
3	Associated service: hearing impaired (HI)
4	Associated service: dialogue (D)
5	Associated service: commentary (C)
6	Associated service: emergency (E)
7	Associated service: voice-over (VO)

- The designation of streams for specific languages is done at the authoring stage instead of selecting from a table as in ATSC. The order of use of the streams designated 0–7 is determined by the author. The language code bytes in the Dolby Digital bit stream are ignored.
- DVD-V in its Dolby Digital metadata streams follows the conventions of the ATSC including Audio Coding Mode with LFE flag, dialnorm, DRC, Mixlevel, Room Type, and Downmixing of center and surround into left and right for 2-channel presentation, and possible Lt/Rt encoding.
- A Karaoke mode, mostly relevant to Asian market players, is supported which permits variable mixing of mono vocal with stereo background and melody tracks in the player.

LPCM is mandatory for all players, and is required on discs that do not have Dolby Digital or MPEG-Layer 2 tracks. Dolby Digital is mandatory for NTSC discs that do not have LPCM tracks; MPEG-2 or Dolby Digital is mandatory for PAL discs that do not have LPCM tracks. Players follow the convention of their region, although in practice Dolby Digital coded discs dominate in much of the world.

Multiple Streams

For the best flexibility in transmission or distribution to cover a number of different audience needs, more than one audio service may be broadcast or recorded. A single audio service may be the complete program to be heard by the listener, or it may be a service meant to be combined with one other service to make a complete presentation. Although the idea of combining two services together into one program is prominent in ATSC documentation, in fact, it is not a requirement of DTV sets to decode multiple streams, nor of DVD players. There are two types of main service and six types of associated services.

Visually Impaired (VI) service is a descriptive narration mono channel. It could be mixed upon reproduction into a CM program, or guidelines foresee the possibility of reproducing it over open-air-type headphones to a VI listener among normally sighted ones. In the case of mixing the services CM and VI together, a gain-control function is exercised by the VI service over the level of the CM service, allowing the VI service provider to "duck" the level of the main program for the description. The Hearing Impaired (HI) channel is intended for a highly compressed version of the speech (dialogue and any narration) of a program. It could be mixed in the end-user's set in proportion with a CM service; this facility was felt to be important as there is not just one kind of hearing impairment or need in this area. Alternatively, the HI channel could be supplied as a separate output from a decoder, for headphone use by the HI listener.

Dialogue service is meant to be mixed with a music and effects (ME) service to produce a complete program. More than one dialogue service could be supplied for multiple languages, and each one could be from mono through 5.1-channel presentations. Further information on multilingual capability is in document A/54: Guide to the Use of the ATSC Digital Television Standard, available from www.atsc.org. Commentary differs from dialogue by being non-essential, and is restricted to a single channel, for instance, narration. The commentary channel acts like VI with respect to level control: the commentary service provider is in charge of the level of the CM program, so may "duck it" under the commentary. Emergency service is given priority in decoding and presentation; it mutes the main services playing when activated. Voice-over (VO) is a monaural, center-channel service generally meant for "voice-overs" at the ends of shows, for instance.

Each elementary stream contains the coded representation of one audio service. Each elementary stream is conveyed by the transport multiplex layer, which also combines the various audio streams with video and with text and other streams, like access control. There are a number of audio service types that may be individually coded into each elementary stream (Table 5-3). Each elementary stream is designated for its service type using a bit field called bsmod (bit stream mode), according to the table above. Each associated service may be tagged in the transport data as being associated with one or more main audio services. Each elementary stream may also be given a language code.

Table 5-3 Typical Audio Bit Rates for Dolby Digital

Type of service (see Table 5-1)	*Number of channels*	*Typical bit rates (kbps)*
CM, ME	5	320–384
CM, ME	4	256–384
CM, ME	3	192–320
CM, ME	2	128–356
VI, narrative only	1	48–128
HI, narrative only	1	48–96
D	1	64–128
D	2	96–192
C, commentary only	1	32–128
E	1	32–128
VO	1	64–128

Three Level-Setting Mechanisms

Dialnorm

Dialnorm is a setting of the audio encoder for the average level of dialogue within a program. The use of dialnorm within a system adopts a "floating reference level" that is based not on an arbitrary level tone, to which program material may only be loosely correlated at best, but instead on the program element that is most often used by people to judge the loudness of a program, namely, the level of speech. Arguments over whether to use −20 or −12 dBFS as a reference are superseded with this new system as the reference level floats from program source to program source, and the receiver or decoder takes action based on the value of dialnorm. Setting dialnorm at the encoder correctly is vitally important, as it is required by the FCC to be decoded and used by receivers to set their gain. There are very few absolute requirements on television set manufacturers, but respecting dialnorm is one of them.

Let us say that a program is news, with a live, on-screen reporter. The average audio level of the reporter is −15 dBFS, that is, the long-term average of the speech is 15 dB below full scale, leaving 15 dB of headroom above the average level to accommodate instantaneous peaks. Dialnorm is set to −15 dB. The next program up is a Hollywood movie. Now, more headroom is needed since there may be sound effects that are much louder than dialogue. The average level of dialogue is −27 dBFS, and dialnorm is set to this value. The following gain adjustments then take place in the television set: during the newsreader the gain is set by dialnorm, and the volume control is set by the user, for a normal level of speech in his listening room, which will probably average about 65 dB SPL for a cross-section of listeners. Dialnorm in this case turns the gain down from −15 dB to −31 dB, a total of 16 dB. Next, a movie comes one, and dialnorm turns the gain down from −27 dB to −31 dB, a total of 4 dB. The difference between the dialogue of the movie and the newsreader of 12 dB has been normalized so both play back at the same acoustic level.

With this system, the best use of the dynamic range of coding is made by each piece of program material, because there is no headroom left unused for the program with the lower peak reproduction level, so no "undermodulation" occurs. The peak levels for both programs are somewhere near the full scale of the medium. Also, interchangeability across programs and channels is good, because the important dialogue cue is standardized, yet the full dynamic range of the programs is still available to end listeners. NTSC audio broadcasting, and CD production too, often achieve interchangeability of product by using a great deal of audio compression so that loudness remains constant across programs, but this restriction on dynamic range makes all programs rather too interchangeable, namely, bland. Proper use of dialnorm prevents this problem.

Dialnorm is the average level of dialogue compared to digital full scale. Such a measurement is called Leq(A), which involves averaging the speech level over the whole length of the program and weighting it according to the A weighting standard curve. "A" weighting is an equalizer that has most sensitivity at mid-frequencies, with decreasing sensitivity towards lower and higher frequencies, thus accounting in part for human hearing's response versus frequency. Meters are available that measure Leq(A). The measurement is then compared to what the medium would be capable of at full scale, and referenced in minus deciBels relative to full scale. For instance, if dialogue measures 76 dB Leq(A) and the full scale 0 dBFS value corresponds to 105 dB SPL (as are each typical for film mixes), then dialnorm is −27 dB.

Applying a measure based on the level of dialogue of course does not work when the program material is purely music. In this case, it is important to match the perceived level of the music against the level of program containing dialogue. Since music may use a great deal of compression, and is likely to be more constant than dialogue, a correction or offset of dialnorm of about 6 dB may be appropriate to match the perceived loudness of the program.

Dialnorm, as established by the ATSC, is currently to the Leq(A) measurement standard. One of the reasons for this was that this method already appeared in national and international standards, and there was equipment already on the market to measure it. On the other hand, Leq(A) does not factor in a number of items that are known to influence the perception of loudness, such as a more precise weighting curve than A weighting, any measure of spectrum density, or a tone correction factor, all of which are known to influence loudness. Still, it is a fairly good measure because what is being compared from one dialnorm measurement to the next is the level of dialogue, which does not vary as much as the wide range of program material. Typical values of dialnorm are shown in Table 5-4.

Table 5-4 Typical Dialnorm Values

Type of program	*Leq(A) (dBFS)*	*Correction (dB)*	*Typical dialnorm (dBFS)*
Sitcom	−18	0	−18
TV drama	−20	0	−20
News/public affairs	−15	0	−15
Sports	−22	0	−22
Movie of the week	−27	0	−27
Theatrical film	−27	0	−27
Violin/piano	−12	−6	−18
Jazz/New Age	−16	−6	−22
Alternative Pop/Rock	−4	−6	−10
Aggressive Rock	−6	−6	−12
Gospel/Pop	−24	−6	−30

Recently Leq(A) has come into question as the best weighting filter and method. Various proponents subjected a wide range of meter types to investigation internationally. The winning algorithm was a relatively simple one, for predicting the response to loudness. It is called LKFS, for equivalent level integrated over time, K weighted, and relative to full scale, in decibels. The K weighting is new to standards but was found to best predict human reaction to program loudness: it is a high-pass filter and a high-frequency shelf filter equalizing upward. As time goes by this method will probably displace Leq(A), but with similar definitions of a time-based measurement compared to full scale, the change should not be too great. At least one broadcaster, NHK, is relying on the much more sophisticated full Zwicker loudness meter.

A loudness-based meter reads many decibels below full scale, and does not predict the approach of the program material to overmodulation. For this, true peak meters are necessary: the two functions could be displayed simultaneously on the same scale. There is a problem with normal 48-kHz sampling and reading of true peaks, as the true peak of a signal within the bandwidth could be higher than the sampler sees. For this reason, true peak requires some amount of oversampling, and 8 × oversampling is adequate to get to a very close approximation. In international standards this has been designated dB TP (decibels true peak). Meters are expected to emerge over

Fig. 5-1 CDs often require the user to adjust the volume control, since program varies in loudness; the use of dialnorm makes volume more constant.

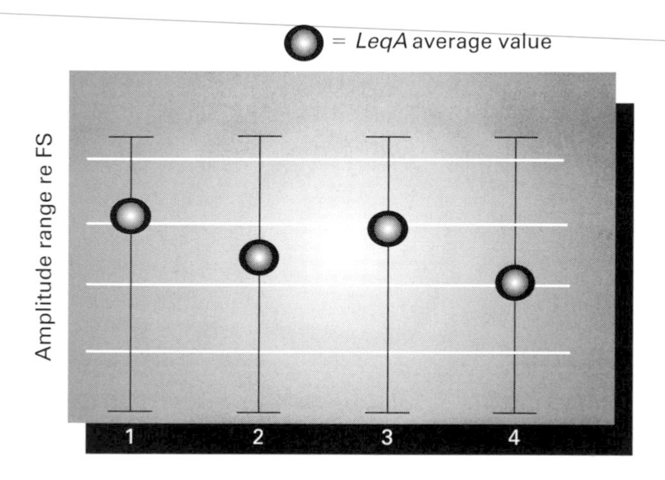

(a) Typical CD production today. Peak levels are adjusted to just reach full scale, but average values vary. Thus user must adjust level for each program.

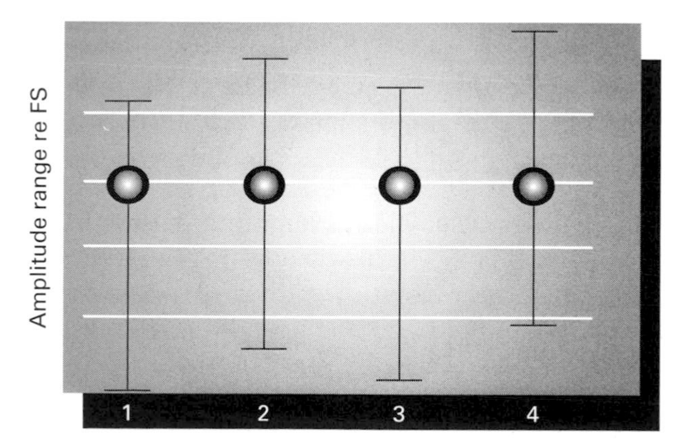

(b) By recording *LeqA* and adjusting it on playback use of dialnorm makes the average level from program to program constant. It does leave a large variation in the maximum SPL.

the next few years that read LKFS for loudness and dB TP, probably simultaneously.

Dynamic Range Compression

While the newsreader of the example above is reproduced at a comfortable 65 dB SPL, and so is the dialogue in the movie, the film has 27 dB of headroom, so its loudest sounds could reach 92 dB SPL (per channel, and 102 dB SPL in the LFE channel). Although this is some 13 dB below the original theatrical level, it could still be too loud at home, particularly at night. Likewise, the softest sounds in a movie are often more than 50 dB below the loudest ones, and could get difficult to hear (Fig. 5-2).

Fig. 5-2 DRC is an audio compression system built into the various coding schemes. Its purpose is to make for a more palatable dynamic range for home listening, while permitting the enthusiast to hear the full dynamic range of the source.

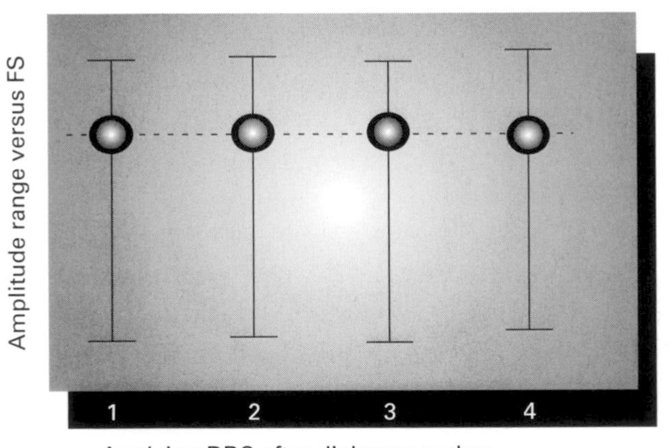

Applying DRC after dialnorm makes the dynamic range of various programs more similar.

Broadcasting has in the past solved this problem by the use of massive compression, used to make the audio level nearly constant throughout programs and commercials. This helped not only consistency within 1 channel, but also consistency in level as the channel is changed. Compressors are used in postproduction, in network facilities, and virtually always at the local station, leading to triple compression being the order of the day on conventional broadcast television. The result is relatively inoffensive but was described above as bland (except there are still complaints about commercials; in one survey of one night promos for network programs stood out even louder than straight commercials).

Since many people may want a restricted dynamic range in reproducing wide dynamic range events like sports or movies at home, a system called Dynamic Range Compression (DRC) has been supplied. Each frame of coded audio includes a "gain word" that sets the amount of compression for that word. To the typical user, the parameters that affect the amount of compression applied, its range, time constants, etc., are controlled by a choice among one of five types of program material: music, music light (compression), film, film light, and speech. Many internal parameters are adjustable, and custom systems to control DRC are under development from some of the traditional suppliers of compression hardware.

Night Listening

The end-user's decoder may have several available possible options: apply DRC all the time, be able to switch DRC on and off (perhaps calling it such names as a Night Switch), or be able to use a variable amount of DRC. This last option permits the end user to apply as little or as much compression as he likes, although making clear to a wide range of end users just what is going on in using a variable amount of the original compression is a challenge. Perhaps a screen display called Audio Compression variable between Full Dynamic Range and Night Time Listening would do the trick.

Mixlevel

Dialnorm and DRC are floating level standards, that is, they do not tie a specific coded value to any particular reproduction sound pressure level. While dialnorm solves interchangeability problems, and DRC dynamic range ones, many psychoacoustic factors are changed in the perception of a program when it is reproduced at a different absolute level than intended by the producer.

An example of the changes that occur accompanying absolute level changes include the equal-loudness effect, wherein listeners perceive less bass as the absolute reproduction level is decreased. This is due to the fact that the equal-loudness contours of human hearing are not parallel curves. That is, although it takes more energy at 100 Hz than at 1 kHz to sound equally loud, this effect varies with level, so that at low levels the amount by which the 100-Hz tone must be turned up to sound as loud as a 1-kHz tone is greater. Thus, in a typical situation where the home listener prefers a lower level than a studio mixer does, the perception of bass is lessened.

Typically, home listeners play programs at least 8–10 dB softer than normal studio listeners. Having an absolute level reference permits home decoders to do a precise job of loudness compensation, that is, best representing the spectrum to the end user despite his hearing it at a lower level. While the "loudness" switch on home stereos has provided some means to do this for years, most such switches are far off the mark of making the correct compensation, due to calibration of sound pressure levels among other problems. Having the mixlevel available solves this problem.

Mixlevel is a 5 bit code representing in 0–31 (decimal) the sound pressure level range from 80 to 111 dB, respectively. The value is set to correspond to 0 dBFS in the digital domain. For film mixes aligned to the 85-dB standard (for −20 dBFS), the maximum level is 105 dB SPL per channel. Mixlevel is thus 25 dB above 80 dB SPL, and should be coded with value 25. Actual hardware will probably put this in terms of reference level for −20 dBFS, or 85 dB SPL for film mixes. Television mixes take place in the range from 78 to 83 dB typically, and music mixes from 80 to 95 dB SPL, all for −20 dBFS in the digital domain.

Audio Production Information Exists

This is a flag that refers to whether the Mixing Level and Room Type information is available.

Room Type

There are two primary types of mixing rooms for the program material reaching television sets: control rooms and Hollywood film-based dubbing stages. These have different electro-acoustic responses according to their size and purpose. Listening in an aligned control room to sound mixed in a Hollywood dubbing stage shows this program material to be not interchangeable. The large-room response is rolled off at high frequencies to the standard SMPTE 202 (ISO 2969). The small room is flatter to a higher frequency, such as in "Listening conditions for the assessment of sound programme material," EBU Tech. 3276-E available from the EBU web site www.edu.ch.

The difference between these two source environments can be made up in a decoder responsive to a pair of bits set for informing the decoder which room type is in use to monitor the program (Table 5-5).

Table 5-5 Room Type

Bit code for roomtyp	*Type of mixing room*
00	Not indicated
01	Large room, X curve monitor
10	Small room, flat monitor
11	Reserved

Dolby Surround Mode Switch

The 2-channel stereo content (2/0) could be from original 2-channel stereo sources, or from Lt/Rt sources used with amplitude-phase 4:2:4 matrixing. Ordinary 2-channel sources produce uncertain results when decoded by way of a matrix decoder, such as Dolby Pro Logic or Pro Logic II. Among the problems could be a reduction in the audible stereo width of a program, or content appearing in the surround loudspeakers that was not intended for reproduction at such a disparate location. On the other hand, playing Dolby Surround or Ultra Stereo encoded movies over 2 channels robs them of the spatial character built into them through the use of center and surround channels.

For these reasons the ATSC system and its derivatives in packaged media employ a flag that tells decoding equipment whether the 2/0 program is amplitude-phase matrix encoded, and thus whether the decoding equipment should switch in a surround decoder such as Pro Logic.

Downmix Options

5.1-channel bit streams are common today, having been used now on thousands of movies, and are increasingly common in digital television. Yet, a great many homes have Pro Logic or Pro Logic II matrix-based receivers. For equipment already sold some years ago it is common for the user's equipment, such as a set-top box, to supply a 2-channel mixdown of the 5.1 channel original. Since program material varies greatly in its compatibility to mixdown, producer options were made a part of the system. Gain constants for mixdown are transmitted in the metadata, for use in the mixdown decoder.

Center channel content is distributed equally into left and right channels of a 2-channel downmix with one of a choice of three levels. Each level is how much of center is mixed into both left and right. The alternatives are −3, −4.5, and −6 dB. The thinking behind these alternatives was as follows:

- −3 dB is the right amount to distribute into two acoustic sources to reach the same sound power level, thus keeping the reverberant field level, as is typical at home, equal. This is the amount by which a standard sin–cos panner redistributes a center panned image into left and right, for instance.
- −6 dB covers the case where the listening is dominated by direct sound. Thus, the two equal source signals add up by 6 dB rather than by 3 dB, because they add as vectors, as voltages do, rather than by 3 dB as power does.
- Since −3 and −6 dB represent the extreme limits (of power addition on the one hand, or of phase-dependent vector addition on the

other), an intermediate, compromise value was seen as valuable, since the correct answer has to be −4.5 dB ± 1.5 dB.

What was not considered by the ATSC in setting this standard is that the center build-up of discrete tracks mixed together in the mixdown process and decoded through an amplitude-phase matrix could cause dialogue intelligibility problems, due to the "pile up" of signals in the center of the stereo sound field. On some titles, while the discrete 5.1-channel source mix has good intelligibility, after undergoing the auto-mixdown to 2-channel Lt/Rt in a set-top box, and decoding in a Pro Logic receiver, even with the center mixdown level set to the highest value of −3 dB, dialogue intelligibility is pushed just over the edge, as competing sound effects and music pile up in the center along with the dialogue.

In such cases, the solution is to raise slightly the level of the center channel in the discrete 5.1-channel mix, probably by no more than 1 to 2 dB. This means that important transfers must be monitored both in discrete, and in matrix downmixes, to be certain of results applicable to most listeners.

Surround downmix level is the amount of left surround to mix into left, and right surround to right, when mixing down from any surround equipped format to 2 channel. The available options are −3 dB, −6 dB, and off. The thinking behind these are as follows:

- −3 dB is the amount by which mono surround information, from many movie mixes before discrete 5.1 was available, mixes down to maintain the same level as the original.
- −6 dB is an amount that makes the mixdown of surround content not so prominent, based on the fact that most surround content is not as important as a lot of front content. This helps to avoid competition with dialogue, for instance, by heavy surround tracks in a mixdown situation.
- Off was felt necessary for cases where the surround levels are so high that they compete with the front channels too much in mixdown situations. For instance, a digital television found on a kitchen counter and a surround mix of football intended for it should not contain crowd sound to the same extent as the large-scale media room presentation.

Level Adjustment of Film Mixes

The calibration methods of motion picture theaters and home theaters are different. In the motion picture theater each of the two surround monitor levels is set 3 dB lower than the front channels, so that their sum adds up to one screen channel. In the home, all 5 channels are set to equal level. This means that a mix intended for theatrical release must be adjusted downwards by 3 dB in each of the surround channels in the transfer to home media. The Dolby Digital encoder software provides for this required level shift by checking the appropriate box.

Lip-Sync and Other Sync Problems

There are many potential sources for audio-to-video synchronization problems. The BBC's "Technical Requirements for Digital Television Services" calls for a "sound-to-vision synchronization" of ±20 ms, 1/2 frame at PAL rate of 25 frames/s. Many people may not notice an error of 1 frame, and virtually everyone is bothered by a 2 frame error. Many times, sync on separate media is checked by summing the audio from the video source tape with the audio from a double-system tape source, such as a DTRS (DA-98 for instance) tape running along in sync, and listening for "phasing" which indicates that sync is quite close. Phasing is an effect where comb filters are produced due to time offset between two summed sources; it is the comb filter that is audible, not the phase per se between the sources. Another way to check sync is not to sum, but rather to play the center channel from the videotape source into the left monitor loudspeaker and the center channel from the double-system source into the right loudspeaker and listen for a phantom image between left and right while sitting exactly on the centerline of a gain matched system, as listeners are very sensitive to time of arrival errors under these conditions. Both these methods assume that there is a conformed audio scratch track that is in sync on the videotape master, and that we are checking a corresponding double-system tape.

The sources for errors include the following, any of which could apply to any given case:

- Original element out of sync.
- ±1/2 frame sync tolerance due to audio versus video framing of Dolby Digital, something that Dolby E is supposed to prevent.
- Improper time stamping during encoding.
- In an attempt to produce synchronized picture and sound quickly, players fill up internal buffers, search for rough sync, then synchronize and keep the same picture–sound relationship going forward, which may be wrong—such players may sync properly by pushing pause, then play, giving an opportunity to refill the buffers and resynchronize.
- If Dolby E is used with digital video recorders, the audio is in sync with the video on the tape, but the audio encoding delay is one video frame and decoding delay is one frame. This means that audio must be advanced by one frame in layback recording situations, to account for the one frame encoding delay. Also, on the output the video needs to be delayed by one frame so that it remains in sync with

the audio, and this requirement applies to both standard and high-definition formats. Dolby E equipment uses 2 channels of the 4 available on the recorder, and the other 2 are still available for LPCM recording, but synchronization problems could occur, so the Dolby E encoder also processes the LPCM channels to incorporate the delays as for the Dolby E process, so sync is maintained between both kinds of tracks. The name for audio delay of the PCM tracks is Utility Delay. Note that the LPCM track on a second medium must also then be advanced by the one frame delay of the Dolby E encoder, even though it is not being encoded.

Since chase lock synchronizers based on SMPTE/EBU time code is at best only precise to about ±20 audio samples, the audio phasing or imaging tests described above are of limited use in the situation where the two pairs of audio channels originate on different source tapes, locked to the video machine through time code. In these cases, it is best to keep the Lt/Rt 2-channel version on the same tape as the coded version, so that sample lock can be maintained, and then to lay back both simultaneously. Failing this, the sync can be checked by playing one pair of tracks panned to the front loudspeakers, and the other pair panned to the surround loudspeakers, and noting that the sync relationship stays correctly constant throughout the program.

Reel Edits or Joins

Another difficulty in synchronization is maintaining sync across source reels, when the output of multiple tapes must be joined together in the mastering process to make a full length program. Low-bit-rate coders like Dolby Digital produce output data streams that are generally not meant to be edited, so standard techniques like crossfading do not work. Instead, once the source tape content has been captured to a computer file, an operator today has to edit the file at the level of hex code to make reel edits. Thus, anyone supplying multireel format projects for encoding should interchange information with the encoding personnel about how the reel edits will be accomplished.

Media Specifics

Film sound started multichannel, with a carryover from 70 mm release print practice to digital sound on film. In a 1987 Society of Motion Picture and Television Engineers (SMPTE) subcommittee, the outline of 5.1-channel sound for film was produced, including sample rate, word length, and number of audio channels. All such systems designed to accommodate the space available on the film had to make use of low-bit-rate coding. By 1991, the first digital sound on film format, Cinema

Digital Sound, was introduced. It failed, due at least in part to the lack of a "back-up" analog track on the film. Then in 1992, with *Batman Returns*, Dolby Digital was introduced, and in 1993 *Jurassic Park* introduced DTS, with a time code on film and a CD-ROM disc follower containing low-bit-rate coded audio. These were joined in 1994 by Sony Dynamic Digital Sound (SDDS), with up to 7.1 channels of capacity. These three coding–recording methods are prominent today for the theatrical distribution environment, although SDDS is in such a phase where no new equipment is being developed. The methods of coding developed for sound related to film subsequently affected the digital television standardization process, and packaged media introductions of Laser Disc and DVD-V.

While the work went on in the ATSC to determine requirements for the broadcast system, and many standards came out of that process (e.g., see A/52–A/54 of the ATSC at www.atsc.org), the first medium for multichannel sound released for the home environment was Laser Disc. Within a fairly short period of time the DVD was introduced, and the era of discrete multichannel audio for the home became prominent. A universal method of transport was developed to send compressed audio over an S/PDIF connection (IEC 958), called "IEC 61937-1 Ed. 1.0 Interfaces for non-LPCM encoded audio bitstreams applying IEC 60958—Part 1: Non-linear PCM encoded audio bitstreams for consumer applications" for both Dolby Digital and DTS coding methods. While Laser Disc players only had room for one coded signal that could carry from 1 to 5.1 channels of audio, DVDs could have a variety of languages and 2- and 5.1-channel mixes. The actual audio on any particular disc depends on the producers "bit budget," affected by program length, video quality, other services such as subtitling.

Digital Versatile Disc

The Digital Versatile Disc is a set of standards that include Video, Audio, and ROM playback-only discs, as well as writable and re-writable versions. The audio capabilities of the Video Disc are given here; those for DVD-A are in Appendix 3. DVD has about seven times the storage capacity of the Compact Disc, and that is only accounting for a one-sided, one-layer disc. Discs can be made dual-layer, dual-sided, or a mixture, for a range of storage capacities. The playback-only (read-only) discs generally have somewhat higher storage capacity than the writable or re-writable discs in the series. The capacities for the play-only discs are given in Table 5-7.

In contrast, the CD has 650-MB capacity, one-seventh that of a single-sided single-layer DVD. The greater capacity of the DVD is achieved through a combination of smaller pits, closer track "pitch" (a tighter spiral), and better digital coding for the media and error correction.

Table 5-6 Capacity of Some Packaged Release Media

Medium	*Method*	*Number of channels/ stream*	*Maximum number of digital streams*	*Metadata*	*Bit rate(s), bps*	
VHS	2 linear +2 "Hi Fi" analog Dolby Stereo; Ultra Stereo	2, Lo/Ro	0	NA	NA	
		2, Lt/Rt for matrix decoding to 4	0	NA	NA	
CD	LPCM	2, Lo/Ro, or rarely, Lt/Rt for decoding to 4	1	No	1.411 Mbps	
	DTS	1–5.1	1	Some	1.411 Mbps	
DVD-A	LPCM	1–6	1 typ.	SMART	9.6 Mbps	See Appendix 3
	LPCM+MLP	1–6	1 typ.	SMART+ extensions		
	Dolby Digital	1–5.1	8	Yes	Up to 448 kbps	
	DTS	1–5.1	7	Some	192–1.536 k/stream	
SACD	Direct Stream Digital	1–6	1	Some	2.8224 Mbps/ch	
Hybrid Disc	SACD layer; CD layer	1–6 on high-density layer; 2 on CD layer	1	Some; none	2.8224 Mbps/ch; 1.411 Mbps	

Table 5-7 DVD Types and Their Capacity*

DVD type	*Number of sides*	*Layers*	*Capacity*
DVD-5	1	1	4.7 GB
DVD-9	1	2	8.5 GB
DVD-10	2	1	9.4 GB
DVD-14	2	1 on 1 side; 2 on opposite side	13.2 GB
DVD-18	2	2	17.0 GB

*Note that the capacity is quoted in DVD in billions of bytes, whereas when quoting file sizes and computer memory the computer industry uses GBytes, which might seem superficially to be identical, but they are not. The difference is that in computers units are counted by increments of 1,024 instead of 1,000. This is yet another peculiarity, like starting the count of items at zero (zero is the first one), that the computer industry uses due to the binary nature of counting in zeros and ones. The result is an adjustment that has to be made at each increment of 1,000, that is, at Kilo, Mega, and Giga. The adjustment at Giga is $1{,}000/1{,}024 \times 1{,}000/1{,}024 \times 1{,}000/1{,}024 = 0.9313$. Thus, the DVD-5 disc, with a capacity of 4.7×10^9 bytes, has a capacity of 4.38 GB, in computer memory terms. Both may be designated GB, although the IEC has recommended since 1999 that the unit associated with counting by 1,024 be called gibibyte, abbreviated GiB. Every operating system uses this for file sizes, and typically for hard disc sizes. Note that 1 byte equals 8 bits under all circumstances, but the word length varies in digital audio, typically from 16 to 24 bits.

Since the pits are smaller, a shorter wavelength laser diode must be used to read them, and the tracking and focus servos must track finer features. Thus, a DVD will not play at all in a CD player. A CD can be read by a DVD player, but some DVD players will not read CD-R discs or other lower than normal reflectance discs. Within the DVD family, not all discs are playable in all players either: see the specific descriptions below.

Audio on DVD-Video

On DVD-V there are from 1 to 8 audio streams (*note*: not channels) possible. Each of these streams can be coded and named at the time of authoring, such as English, French, German, and Director's Commentary. The order of the streams affects the order of presentation from some players that typically default to playing stream 1 first. For instance, if 2-channel Dolby Surround is encoded in stream 1, players will default to that setting. To get 5.1-channel discrete sound, the user will have to switch to stream 2. The reason that some discs are made this way is that a large installed base of receivers is equipped with Pro Logic decoding, so that the choice of the first stream satisfies this large market. On the other hand, it makes users who want discrete sound have to take action to get it. DVD players can generally switch among the 8 streams, although cannot add streams together. A variety of number of channels and coding schemes can be used, making the DVD live up to the versatile part of its name. Table 5-8 shows the options available. Note that the number of audio streams and their coding options must be traded off against picture quality. DVD-V has a maximum bit rate of 10.08 Mbps for the video and all audio streams. The actual rate off the disc is higher, but the additional bit rate is used for the overhead of the system, such as coding for the medium, error coding, etc. The video is usually encoded with a variable bit rate, and good picture quality can often be achieved using an average of as little as 4.5 Mbps. Also, the maximum bit rate for audio streams is given in the table at 6.144 Mbps, so all of the available bit rate cannot be used for audio only. Thus, one tradeoff that might be made for audio accompanying video is to produce 448-kbps multichannel sound with Dolby Digital coding for the principal language, but provide an Lt/Rt at a lower bit rate for secondary languages, among the 8 streams.

A complication is: "How am I going to get the channels out of the player?" If a player is equipped with six analog outputs, there is very little equipment in the marketplace that accepts six analog inputs, so there is little to which the player may be connected. Most players come equipped with two analog outputs as a consequence, and multichannel mixes are downmixed internally for presentation at these outputs.

Table 5-8 Audio Portion of a DVD-V Disc

Audio coding method	*Sample rate (kHz)*	*Word length*	*Maximum number of channels*	*Bit rates*
LPCM	48	16	8	Maximum 6.144 Mbps
	48	20	6	
	48	24	4	
	96	16	4	
	96	20	3	
	96	24	2	
Dolby Digital	48	up to 24	6	32–448 kbps/stream
MPEG-2	48	16	8	Maximum 912 kbps/stream
DTS	48	up to 24	6	192 k–1.536 Mbps/stream

If more than 2 channels of 48 kHz LPCM are used, the high bit rates preclude sending them over a single-conductor digital interface. Dolby Digital or DTS may be sent by one wire out of a player on S/PDIF format standard modified so that following equipment knows that the signal is not LPCM 2 channel, but instead multichannel coded audio (per IEC 61937). This then is the principal format used to deliver multichannel audio out of a DVD player and into a home sound system, with the Dolby Digital or DTS decoder in the receiver. The connection may either be coaxial S/PDIF, or optical, usually TOSLINK.

In addition, there are a large number of subtitling language options that are outside the scope of this book.

All in all, you can think of DVD as a "bit bucket" having a certain size of storage, and a certain bit rate out of storage that are both limitations defining the medium. DVD uses a file structure called Universal Disc Format (UDF) developed for optical media after a great deal of confusion developed in the CD-ROM business with many different file formats on different operating systems. UDF allows for Macintosh, UNIX, Windows, and DOS operating systems as well as a custom system built into DVD players to read the discs. A dedicated player addresses only the file structure elements that it needs for steering, and all other files remain invisible to it. Since the file system is already built into the format for multiple operating systems, it is expected that rapid adoption will occur in computer markets.

There are the following advantages and disadvantages of treating the DVD-V disc as a carrier for audio "mostly" program. (There might be accompanying still pictures, for instance.)

- *Installed base of players*: Audio on a DVD-V can play on all the DVD players in the field, whereas DVD-A releases will not play on DVD-V players, unless they carry an optional Dolby Digital track in a video partition.
- *No confusion in the marketplace*: DVD is a household name; however, to differentiate it between video and audio sub-branches is beyond the capacity of the marketplace to absorb, except at the very high end, for the foreseeable future.
- DVD-V is not as flexible in its audio capabilities as the DVD-A format.
- 96-kHz/24-bit LPCM audio is only available on 2 channels, which some players down sample (decimate) to 48 kHz by skipping every other sample and truncating at 16 bits; thus uncertain quality results from pressing "the limits of the envelope" in this medium.

HD DVD and Blu-Ray Discs

With the large success of the DVD about 10 years old now, demand for in particular better picture quality is thought by manufacturers to exist. Two camps, HD DVD and Blu-ray, have emerged and it is not clear whether one or both will survive in the long run, especially in light of the beginning of Internet downloads. The audio capabilities are generally limited to 8 channels of audio per stream, with a large variety of coding systems available, as shown in Table 5-9.

In addition, there are differences in some additional constraints between the HD DVD and Blu-ray standards that are illuminated in the two standards.[3] The interactive features built into the standards for each of these have not been employed much as of yet, but there is expected to be expanding capabilities over time with these formats, especially with players connected to the Internet.

Digital Terrestrial and Satellite Broadcast

ATSC set the standard in the early 1990s, however it took some time to get on the air. At this writing, although postponed in the past, NTSC is scheduled to be shut off February 17, 2009. From 2005 testimony before

---

[3]http://www.dvdfllc.co.jp/pdf/bookcon2007-04.pdf
http://www.Bluraydisc.com/assets/downloadablefile/2b_bdrom_audiovisualapplication_0305-12955.pdf

Table 5-9 Capacity of Disc Release Media Carrying Audio and Video

Audio format name	***Capabilities***	***DVD***	***HD DVD***	***Blu ray***
LPCM		Either LPCM or Dolby Digital required	Mandatory of players; up to 5.1 channels	Mandatory of players; up to 8 channels
Dolby Digital (DD)	2.0 or 5.1 channel with potential for typically one additional channel* via matrixed Dolby Surround EX; up to 640 kbps but see disc limitations; lossy coder	Either Dolby Digital or LPCM required; up to 448 kbps	Mandatory of players; up to 448 kbps	Mandatory of players; up to 640 kbps
Dolby Digital Plus (DD+)	7.1 channels and beyond, limited by players to 8 total channels; up to 6 Mbps; supported by HDMI interface standard; lossy coder	NA	Mandatory of players; up to 3 Mbps	Optional of players <1.7 Mbps
Dolby TrueHD	Up to 8 channels; up to 18 Mbps; supported by HDMI interface; lossless coder	NA	Mandatory of players; optional for discs	Mandatory of players; optional for discs
DTS	Up to 5.1 channels at 192-kHz sampling and 7.1 channels at 96-kHz sampling	Optional for players; optional for discs	Core 5.1 decoding mandatory of players; optional for discs	Core 5.1 decoding mandatory of players; optional for discs
DTS-HD	Up to 8,192 channels but limited to 8 channels on HD DVD and Blu ray; bit rates as given; supported by HDMI 1.1 and 1.2 interfaces; lossy coder; if with constant bit rate >1.5 Mbps name is DTS-HD High Resolution	NA	Core 5.1 decoding mandatory of players; extensions to higher sample rates and channels over HDMI; or optionally complete decoding in player ≥3.0195 Mbps	Core 5.1 decoding mandatory of players; extensions to higher sample rates and channels over optional HDMI; or optional complete decoding in player ≥6.036 Mbps
DTS-HD Master Audio	Up to 8,192 channels and 384-kHz sampling, but limited to 8 channels and 96-kHz sampling on HD DVD and Blu ray; bit rates as given; supported by HDMI 1.3 interface; lossless coder	NA	Core 5.1 decoding mandatory of players; extensions to higher sample rates and channels over HDMI; or optionally complete decoding in player Variable bit rate; peak rate ≥18.432 Mbps	Core 5.1 decoding mandatory of players; extensions to higher sample rates and channels over optional HDMI; or optional complete decoding in player Variable bit rate; peak rate ≥25.524 Mbps

*The possibility exists to use a 4:2:4 matrix on the LS and RS recorded channels. The Left/Back/Right surround is the standard use.

Congress, between 17 and 21 million of approximately 110 million television-equipped households rely on over-the-air broadcasts, disproportionately represented by minority households. Nevertheless, the shut off date is now "harder" than it was in the past. The technical reason for the changeover is the duplication of resources when both analog and digital broadcasts are made in parallel, and the better spectrum utilization of digital compared to analog broadcasts. The spectrum of a DTV channel, although the same bandwidth (6 MHz) as an NTSC one, is more densely packed. The transition has been faster than that to color, showing good consumer acceptance. Simple set-top boxes may be supported by Congress for those left behind in the transition.

All of the features described above were developed through the standards process, and terrestrial television has the capability to support them. However, set manufacturers have likely chosen not to implement certain features, such as needing two stream decoders so that some of the set-mixing features could be implemented. Thus these features remain documented but not implemented. One feature that is required of all receivers is dialnorm. In fact, a high-definition television set is not required to produce a picture, but it is required to respect dialnorm!

Satellite broadcast follows the ATSC largely. However, due to some restrictions on bandwidth it is likely that only single CM programs are broadcast. At this writing, competition among suppliers is increasing, with potential sources being over-the-air, satellite dish, cable set-top boxes, CableCard plugged into "digital cable ready" sets, and what have traditionally been phone companies who have increased the bandwidth of their infrastructure to handle Internet and video signals. These signals are RF for over-the-air and satellite broadcast, and either wideband copper or even fibre optic to the home for the others.

Downloadable Internet Connections

Legal movie downloads have now begun, although they represent only a tiny fraction of multichannel listening so far. The audio codecs used are potentially wide ranging, although Windows Media 9 seems to dominate legal services at this writing. Since this is a low-bit-rate coder it should not be cascaded with a Dolby Digital AC-3 coder, so studios must supply at a minimum Dolby E coded audio for ingest into the servers of these services.

Video Games

Popular video games follow the 2.0- or 5.1-channel standards generally speaking, with typically Dolby Digital or DTS coding at their consumer accessible port. Since video games must deliver sound "on the fly," and since sound is late in the project schedule and occurs when most of the budget has already been expended, sound quality varies enormously, from really pretty bad, to approaching feature film sound design with positional verisimilitude and immersive ambiences, albeit with some limitations due to the "live" generation of material.

While Dolby Digital and DTS provide for a simple connection into a home theater by one S/PDIF connection, they are not generally suitable for internal representations of audio since they would require decoding, summing with other signals, signal processing, and re-encoding to make a useful output. Thus internal systems predominate, in some cases proprietary to the game manufacturers and in other cases by use of operating system component parts. Most internal representations are LPCM, in AIFF or .wav formats for simplicity in signal processing.

This is a rapidly changing field and it is beyond the scope of this book to describe computer game audio any further. Contemporaneous web site information will be of more use to you. For instance, Microsoft's Direct Sound3D which has been around since 1997 is slated to be replaced by XACT and Xaudio 2 with enhanced surround support in late 2007 due to increasing needs for cross platform compatibility between computers and games, among other things.[4]

Digital Cinema

With the coming of digital projection to cinemas has come server-based supply of content, with audio and content protection. The audio standards for digital cinema are under fewer constraints on bit bucket size and bit rate than are those for film because the audio is on a server, and even PCM coded it is still a small fraction of the picture requirements. Also, today hard disc drives are much cheaper than they were even 5 years ago, lessening the requirement for absolute efficiency.

The audio standards call for 48-kHz sampling, with optional 96-kHz sampling that has not been implemented to date, 24-bit LPCM coding (well beyond the dynamic range of any sound system available to play it back on), and up to 16 audio channels. For now the standard 5.1 and quasi-6.1 systems are easily supported, and two of the audio channels may be reserved for HI and VI services, so 14 audio channels are available within the standard, although equipment isn't generally built to such a generous number yet.

[4]The Microsoft Game Developer's Conference 2007 referred to www.msdn.com/direct x as the source for further information when available as of this writing.

Post-Production Media Label for DTV Multichannel-Audio

Post Production Studio Info

Studio Name______________________________________

Studio Address____________________________________

Studio Phone Number______________________________

Contact Person____________________________________

Date Prepared (e.g., 1999-01-12)____________________________

Program Info

Producing Organization______________________________

Program____

Episode Name________________Episode # (1-4095)________

Version ______________________Version # (1-4095)________

First Air or Street Date (e.g., 1999-01-12)______________________

Program Length (Time)__________________________________

Contents Info

❑ Original Master ❑ Simultaneous Protection Master

❑ Protection Dub If original, does simultaneous protection master exist? ❑ Yes ❑ No

If no, why not?____

Track Layout

Trk #	1	2	3	4	5	6	7	8
	L	R	C	LFE	LS	RS	LT	RT
Other								

Leader Contents

❑ 1 kHz sine wave tone at −20 dBFS

❑ Pink noise at −20 dBFSrms

❑ 2 Pop

❑ Other__________________________________

Program starts at 01:00:00:00 Other______________________

Program ends at _____________________________________

❑ Multiple program segments:

Time Code

❑ 29.97 DF ❑ 29.97 NDF ❑ 30.00 DF ❑ 30.00 NDF

❑ Other ___________

Sample Rate

❑ 48.000 kHz ❑ Other ______________________

Info for Dolby Digital Coder

Bit Stream Mode: ❑ Complete Main ❑ Other: ___________________

Audio Coding Mode: ❑ 3/2L ❑ 2 ❑ Other: _____

Low Frequency Effects Channel: ❑ On ❑ Off

Dialogue Normalization: – ____ dB

Dynamic Range Compression: ❑ On ❑ Off

❑ Film Light ❑ Film Standard

❑ Music Light ❑ Music Standard

❑ Speech

Center Downmix Level: ❑ −3 ❑ −4.5 ❑ −6 dB

Surround Downmix Level: ❑ −3 ❑ −6 dB ❑ Off

Dolby Surround Mode: ❑ On ❑ Off

Audio Production Info Exists: ❑ Yes ❑ No

Mixing Level: _____ Room Type: ❑ Small ❑ Large (X curve)

Multichannel-Audio Post-Production Media Label
for 8-track Digital Media
such as DTRS (DA-88, DA-98)

Postproduction Studio Info

Studio Name ______________________________

Studio Address ______________________________

Studio Phone Number ______________________

Contact Person ____________________________

Date Prepared (e.g., 2008-01-25) ____________________

Program Info

Producing Organization ______________________

Program Title ______________________________

Artist(s) ________________________________

Version ________________________________

Program Length (Time) ________________________

Contents Info

❑ Original Master ❑ Simultaneous Protection Master

❑ Digital Clone ❑ Other ______________

If original, does simultaneous protection master exist?

❑ Yes ❑ No

If original, does digital clone exist? ❑ Yes ❑ No

Time Code

❑ 29.97 DF ❑ 29.97 NDF ❑ 30.00 DF

❑ 30.00 NDF ❑ Other _________

Sample Rate

❑ 48.000 kHz

❑ Other____________________

Track Layout

Track #	1	2	3	4	5	6	7	8
❑	L	R	C	LFE	LS	RS		
❑ Other								

Leader Contents ❑ 1 kHz sine wave tone at −20 dBFS

❑ Pink noise at −20 dBFSrms

❑ Other ___________________________

Program starts at 01:00:00:00 Other ______________________

Program ends at ___________________________________

❑ Multiple program segments:

6 Psychoacoustics

Tips from This Chapter

- Localization of a source by a listener depends on three major effects: the difference in level between the two ears, the difference in time between the two ears, and the complex frequency response caused by the interaction of the sound field with the head and especially the outer ears (head-related transfer functions). Both static and dynamic cues are used in localization.
- The effects of head-related transfer functions of sound incident on the head from different angles call for different equalization when sound sources are panned to the surrounds than when they are panned to the front, if the timbre is to be maintained. Thus, direct sounds panned to the surrounds will probably need a different equalization than if they were panned to the front.
- The minimum audible angle varies around a sphere encompassing our heads, and is best in front and in the horizontal plane, becoming progressively worse to the sides, rear, and above and below.
- Localization is poor at low frequencies and thus common bass subwoofer systems are perceptually valid.
- Low-frequency enhancement (LFE) (the 0.1 channel) is psychoacoustically based, delivering greater headroom in a frequency region where hearing is less sensitive.
- Listeners perceive the location of sound from the first arriving direction typically, but this is modified by a variety of effects due to non-delayed or delayed sound from any particular direction. These effects include timbre changes, localization changes, and spaciousness changes.
- Phantom image stereo is fragile with respect to listening position, and has frequency response anomalies.

- Phantom imaging, despite its problems, works more or less in the quadrants in front and behind the listener, but poorly at the sides in 5-channel situations.
- Localization, spaciousness, and envelopment are defined. Methods to produce such sensations are given in Chapter 4. Lessons from concert hall acoustics are given for reverberation, discrete reflections, directional properties of these, and how they relate to multichannel sound.
- Instruments panned partway between front and surround channels in 5.1-channel sound are subject to image instability and sounding split in two spectrally, so this is not generally a good position to use for primary sources.

Introduction

Psychoacoustics is the field pertaining to perception of sound by human beings. Incorporated within it are the physical interactions that occur between sound fields and the human head, outer ears, and ear canal, and internal mechanisms of both the inner ear transducing sound mechanical energy into electrical nerve impulses and the brain interpreting the signals from the inner ears. The perceptual hearing mechanisms are quite astonishing, able to tell the difference when the sound input to the two ears is shifted by just 10 μs, and able to hear over 10 octaves of frequency range (visible light covers a range of less than one octave) and over a tremendous dynamic range, say a range of 10 million to one in pressure.

Interestingly, in one view of this perceptual world, hearing operates with an ADC in between the outer sound field and the inner representation of sound for the brain. The inner ear transduces mechanical waves on its basilar membrane, caused by the sound energy, into patterns of nerve firings that are perceived by the brain as sound. The nerve firings are essentially digital in nature, while the waves on the basilar membrane are analog.

Whole reference works such as Jens Blauert's *Spatial Hearing* and Brian C. J. Moore's *An Introduction to the Psychology of Hearing*, as well as many journal articles, have been written about the huge variety of effects that affect localization, spaciousness, and other topics of interest. Here we will examine the primary factors that affect multichannel recording and listening.

Principal Localization Mechanisms

Since the frequency range of human hearing is so very large, covering 10 octaves, the human head is either a small appearing object (at low frequencies), or a large one (at high frequencies), compared to the wavelength of the sound waves. At the lowest audible frequencies where the wavelength of sound in air is over 50 ft (15 m), the head appears as a small object and sound waves wrap around the head easily through the process called diffraction. At the highest audible frequencies, the wavelength is less than 1 in. (25 mm), and the head appears as a large object operating more like a barrier than it does at lower frequencies. Although sound still diffracts around the barrier, there is an "acoustic shadow" generated towards one side for sound originating at the opposite side.

The head is an object with dimensions associated with mid-frequency wavelengths with respect to sound, and this tells us the first fundamental story in perception: one mechanism will not do to cover the full range, as things are so different in various frequency ranges. At low frequencies, the difference in level at the two ears from sound originating anywhere is low, because the waves flow around the head so freely; our heads just aren't a very big object to a 50-ft wave. Since the level differences are small, localization ability would be weak if it were based only on level differences, but another mechanism is at work. In the low-frequency range, perception relies on the difference in time of arrival at the two ears to "triangulate" direction. This is called the interaural time difference (ITD). You can easily hear this effect by connecting a 36-in. piece of rubber tubing into your two ears and tapping the tubing. Tapped at the center you will hear the tap centered between your ears, and as you move towards one side, the sound will quickly advance towards that side, caused by the time difference between the two ears.

At high frequencies (which are short wavelengths), the head acts more like a barrier, and thus the level at the two ears differs depending on the angle of arrival of the sound at the head. The difference in level between the two ears is called the interaural level difference (ILD). Meanwhile the time difference becomes less important for if it were great confusion would result. The reason for this is that at short wavelengths like 1 in., just moving your head a bit would affect the localization results strongly, and this would have little purpose.

These two mechanisms, time difference at low frequencies and level difference at high ones, account for a large portion of the ability to perceive sound direction. However, we can still hear the difference in direction for sounds that create identical signals at the two ears, since a sound directly in front of us, directly overhead, or directly behind produce identical ILD and ITD. How then do we distinguish such directions? The pinna or shape and convolutions of the outer ear interact differently for sound coming from various directions, altering the frequency response through a combination of resonances and reflections unique for each direction, which we come to learn as associated with that direction. Among other things, pinna effects help in the perception of height.

The combination of ILD, ITD, and pinna effects together form a complicated set of responses that vary with the angle between the sound field and the listener's head. For instance, a broadband sound source containing many frequencies sounds brightest (i.e., has the most apparent high frequencies) when coming directly from one side, and slightly "darker" and duller in timbre when coming from the front or back. You can hear this effect by playing pink noise out of a single loudspeaker and rotating your head left and right. A complex examination of the frequency and time responses for sound fields in the two ear canals coming from a given direction is called a head-related transfer function (HRTF). A thorough set of HRTFs, representing many angles all around a subject or dummy head in frequency and time responses constitute the mechanism by which sound is localized.

Another important factor is that heads are rarely clamped in place (except in experiments!), so there are both static cues, representing the head fixed in space, and dynamic cues, representing the fact that the head is free to move. Dynamic cues are thought to be used to make unambiguous sound location from the front or back, for instance, and to thus resolve "front–back" confusion.

The Minimum Audible Angle

The minimum audible angle (MAA) that can be discerned by listeners varies around them. The MAA is smallest straight in front in the horizontal plane and is about 1°, whereas vertically it is about 3°. The MAA remains good at angles above the plane of listening in front, but becomes progressively worse towards the sides and back. This feature is the reason that psychoacoustically designed multichannel sound systems employ more front channels than rear ones.

Bass Management and Low-Frequency Enhancement Pyschoacoustics

Localization by human listeners is not equally good at all frequencies. It is much worse at low frequencies, leading to practical satellite–subwoofer systems where the low frequencies from the multiple channels are extracted, summed, and supplied to just one subwoofer. Experimental work sought the most sensitive listener from among a group of professional mixers and then found the most sensitive program material (which proved to be male speech, not music). The experiment varied the crossover frequency from satellite to a displaced subwoofer. From this work, a selection of crossover frequency could be made as two standard deviations below the mean of the experimental result from the most sensitive listener listening to the program material found to be most sensitive: that number is 80 Hz. Many systems are based on this crossover frequency, but professionals may choose monitors that go somewhat lower than this, to 50 or 40 Hz commonly. Even in these cases it is important to re-direct the lowest bass from the multiple channels to the subwoofer in order to hear it; otherwise home listeners with bass management could have a more extended bass response than the professional in the studio, and low-frequency problems could be missed.

The LFE (low-frequency enhancement) channel (the 0.1 of 5.1-channel sound), is a separate channel in the medium from producer to listener. The idea for this channel was generated by the psychoacoustic needs of listeners. Systems that have a flat overload level versus frequency perceptually overload first in the bass. This is because at no level is perception flat: it requires more level at low frequencies to sound equally as loud as in the mid-range. Thus the 0.1 channel, with a bandwidth of 1/400 the sample rate of 44.1 or 48 kHz sampled systems (110 or 120 Hz), was added to the 5 main channels of 5.1-channel systems, so that headroom at low frequencies could be maintained at levels that more closely match perception. The level standards for this channel call for it to have 10 dB greater headroom than any one of the main channels in its frequency band. This channel is monaural, meant for special program material that requires large low-frequency headroom. This may include sound effects, and in some rare instances, music and dialogue. An example of the use of LFE in music is the cannon fire in the 1812 Overture, and for dialogue, the voice of the tomb in *Aladdin*.

The 10 dB greater headroom on the LFE channel is obtained by deliberately recording 10 dB low on the medium and then boosting by 10 dB in the playback electronics after the medium. Obviously with a linear medium the level is reproduced as it went into this pair of offsets, but the headroom is increased by 10 dB. Of course, the signal-to-noise ratio is also decreased by 10 dB, but this does not matter because we are speaking of frequencies below 120 Hz where hearing is insensitive to noise. A study by a Dolby Labs engineer of the peak levels on various channels of the 5.1-channel DVD medium found that the recorded level maximum peaks in the LFE channel are about the same as those in the highest of the other 5 channels, which incidentally is the center channel. Since this measurement was made before the introduction of the 10-dB gain after the medium, it showed the utility of the 10-dB offset.

In film sound, the LFE channel drives subwoofers in the theater, and that is the only signal to drive them. In broadcast and packaged video media sound played in the home, LFE is a channel that is usually bass managed by being added together with the low bass from the 5 main channels and supplied to one or more subwoofers.

Effects of the Localization Mechanisms on 5.1-Channel Sound

Sound originating at the surrounds is subject to having a different timbre than sound from the front, even with perfectly matched loudspeakers, due to the effects of the differing HTRFs between the angles of front and surround channels.

In natural hearing, the frequency response caused by the HRTFs is at least partially subtracted out by perception, which uses the HRTFs in the localization process but then more deeply in perception discovers the "source timbre," which remains unchanging with angle. An example is that of a violin played by a moving musician. Although the transfer function (complex frequency response) changes dramatically as the musician moves around a room due to both the room acoustic differences between point of origin and point of reception, and the HRTFs, the violin still sounds like the same violin to us, and we could easily identify a change if the musician picked up a different violin. This is a remarkable ability, able to "cut through" all the differences due to acoustics and HRTFs to find the "true source timbre." This effect, studied by Arthur Benade among others, could lead one to conclude that no equalization is necessary for sound coming from other directions than front, that is, matched loudspeakers and room acoustics, with room equalization performed on the monitor system, might be all that is needed. In other words, panning should result in the same timbre all around, but it does not. We hear various effects:

- For sound panned to surrounds, we perceive a different frequency response than the fronts, one that is characterized by being brighter.
- For sound panned halfway between front and surrounds, we perceive some of the spectrum as emphasized from the front, and other parts from the surround—the sound "tears in two" spectrally and we hear two events, not a single coherent one between the loudspeakers.
- As a sound is panned dynamically from a front speaker to a surround speaker, we hear first the signal split in two spectrally, then come back together as the pan is completed.

All these effects are due to the HRTFs. Why doesn't the theory of timbre constancy with direction hold for multichannel sound, as it does in the case of the violinist? The problem with multichannel sound is that there are so few directions representing a real sound field that a jumpiness between channels reveals that the sound field is not natural. Another way to look at this is that with a 5.1-channel sound system we have coarsely quantized spatial direction, and the steps in between are audible.

The bottom line of this esoteric discussion is: it is all right to equalize instruments panned to the surrounds so they sound good, and that equalization is likely to be different from what you might apply if the instrument is in front. This equalization is complicated by the fact that front loudspeakers produce a direct sound field, reflected sound, and reverberated sound, and so do surround loudspeakers, albeit with different responses. Different directivity loudspeakers interacting with different room acoustics have effects as the balance among these factors vary too. In the end, the advice that can be given is that in all likelihood there will be high-frequency dips needed in the equalization of program sources panned to the surrounds to get it to sound correct compared to frontal presentation. The anechoic direct-sound part of this response is shown in Fig. 6-1.

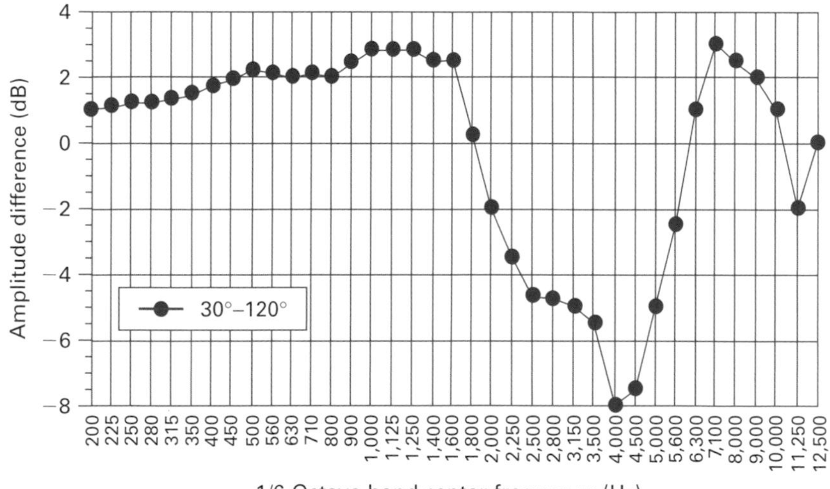

Fig. 6-1 The frequency response difference of the direct sound for a reference loudspeaker located at 30° to the right of straight ahead in the conventional stereo position to one located at 120° away from straight ahead, measured in the ear canal. This would be the equalization to apply to the right surround to get it to match the front right channel for direct sound, but not for reflections or reverberation. Thus this curve is not likely to be directly useful as an equalizer, but it shows that you should not be adverse to trying equalization to better match instrument timbre pannel to surround. Data from E. A. G. Shaw, "Transformation of Sound Pressure Level from the Free Field to the Eardrum in the Horizontal Plane," *J. Acoust. Soc. Am.* Vol. 56, No. 6, pp. 1848–1861.

The Law of the First Wavefront

Sound typically localizes for listeners to the direction of the first source of that sound to arrive at them. This is why we can easily localize sound in a reverberant room, despite considerable "acoustic clutter" that would confuse most technical equipment. For sound identical in level and spectrum supplied by two sources, a phantom image may be formed with certain properties discussed in the next section. In some cases, if later arriving sound is at a higher level than the first, then a phantom image may still be formed. In either of these cases a process called "summing localization" comes into play.

Generally, as reflections from various directions are added to direct sound from one direction, a variety of effects occur. First, there is a sensation that "something has changed," at quite a low threshold. Then, as the level of the reflection becomes higher, a level is reached where the source seems broadened, and the timbre is potentially changed. At even higher levels of reflections, summing localization comes into play, which was studied by Haas and that is why his name is brought up in conjunction with the Law of the First Wavefront. In summing localization a direction intermediate between the two sound sources is heard as the source: a phantom image.

For multichannel practitioners, the way that this information may be put to use is primarily in the psychoacoustic effects of panners, described in Chapter 4, and in how the returns of time delay and reverberation devices are spread out among the channels. This is discussed below under the section "Localization, Spaciousness, and Envelopment."

Phantom Image Stereo

Summing localization is made use of in stereo and multichannel sound systems to produce sound images that lie between the loudspeaker positions. In the 2-channel case, a centered phantom is heard by those with normal hearing when identical sound fields are produced by the left and right loudspeakers, the room acoustics match, and the listener is seated on the centerline facing the loudspeakers.

There are two problems with such a phantom image. The first of these is due to the Law of the First Wavefront: as a listener moves back and forth, a centered phantom moves with the listener, snapping rather quickly to the location of the loudspeakers left or right depending on how much the listener has moved to the left or right. One principal rationale for having a center channel loudspeaker is to "throw out an anchor" in the center of the stereo sound field to make moving left and right, or listening from off center positions generally, hear centered content in the center. With three loudspeakers across the front of the stereo sound field in a 5.1-channel system at 0° (straight ahead) and ±30°, the intermediate positions at left-center and right-center are still subject to image pulling as the listening position shifts left and right, but the amount of such image shift is much smaller than in the 2-channel system with 60° between the loudspeakers.

A second flaw of phantom image listening is due to the fact that there are four sound fields to consider for phantoms. In a 2-channel system for instance, the left loudspeaker produces sound at both the left and right ears, and so does the right loudspeaker. The left loudspeaker sound at the right ear can be considered to be crosstalk. A real centered source would produce just one direct sound at each ear, but a phantom source produces two. The left loudspeaker sound at the right ear is slightly delayed (≈200 μs) compared to the right loudspeaker sound, and subject to more diffraction effects as the sound wraps around the head. For a centered phantom, adding two sounds together with a delay and considering the effects of diffraction, leads to a strong dip around 2 kHz, and ripples in the frequency response at higher frequencies.

This dip at 2 kHz is in the presence region. Increases in this region make the sound more present, while dips make it more distant. Many professional microphones have peaks in this region, possibly for the reason that they are routinely evaluated as a soloist mike panned to the center on a 2-channel system. Survival of the fittest has applied to microphone responses here, but in multichannel, with a real center, no such equalization is needed, and flatter microphones may be required.

Phantom Imaging in Quad

Quad was studied by the BBC Research Laboratories thoroughly in 1975. The question being asked was whether broadcasting should adopt a 4-channel format. The only formal listening tests to quadraphonic sound reproduction resulted in the graph shown in Fig. 6-2. The concentric circles represent specific level differences between pairs of channels. The "butterfly" petal drawn on the circular grid gives the position of the sound image resulting from the inter-channel level differences given by the circles. For instance, with zero difference between left and right, a phantom image in center front results, just as you would expect. When the level is 10 dB lower in the right channel than the left, imaging takes place at a little over 22.5° left of center. The length of the line segments that bracket the inter-channel level difference circles gives the standard deviation, and at 22.5° left the standard deviation is small. When the inter-channel level difference reaches 30 dB, the image is heard at the left loudspeaker.

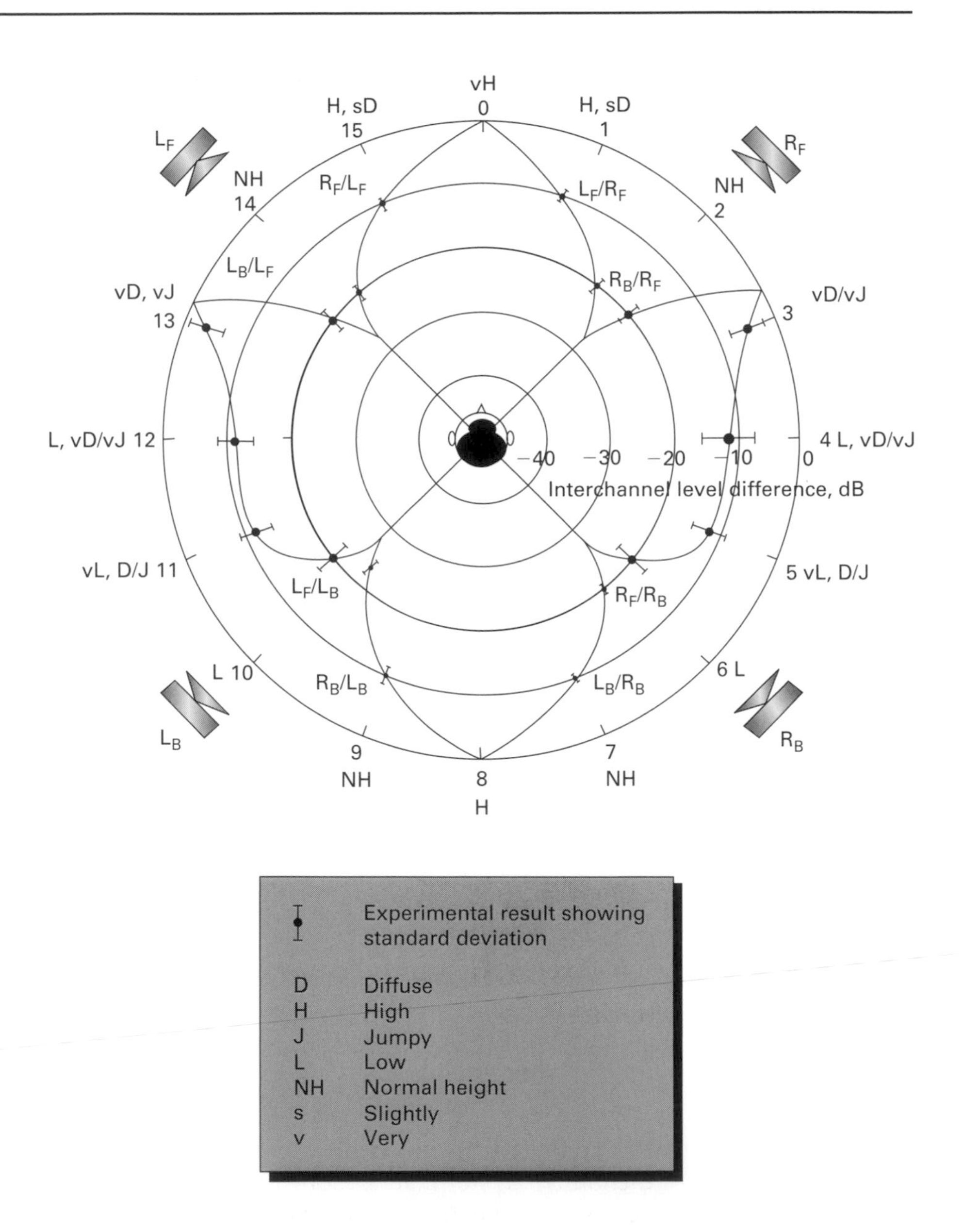

Fig. 6-2
Phantom Imaging in Quad, from BBC Research Reports, 1975.

Now look at the construction of side phantom images. With 0 dB interchannel level difference, the sound image is heard at a position way in front of 90°, about 25° in front of where it should be, in fact. The standard deviation is also much higher than it was across the front, representing differences from person to person. The abbreviations noting the quality of the sound images is important too. The sound occurring where L_B/L_F are equal (at about 13) is labeled vD, vJ, translates to very diffuse and very "jumpy," that means the sound image moves around a lot with small head motions.

Interestingly, the rear phantom image works as well as the center front in this experiment. The reason that the sides work differently from the front and back is of course due to the fact that our heads are not symmetrical with respect to these four sound fields: we have two ears, not four!

Thus, it is often preferred to produce direct sound from just one loudspeaker rather than two, because sound from two produces phantom images that are subject to the precedence effect and frequency response anomalies. This condition is worse at the sides than in the front and rear quadrants. As Blauert puts it: "Quadrophony can transmit information about both the direction of sound incidence and the reverberant sound field. Directions of sound incident in broad parts of the horizontal plan (especially the frontal and rear sections, though not the lateral sectors) are transmitted more or less precisely. However, four loudspeakers and four transmission channels fall far short of synthesizing the sound field at one position in a concert hall faithfully enough so that an attentive listener cannot notice considerable differences in comparison with the original sound field."

Localization, Spaciousness, and Envelopment

The discussion of localization so far has centered on locating a sound at a point in space. Real-world sources may be larger than point sources, and reverberation and diffuse ambiences are meant to be the opposite of localized, that is, diffuse. Some ideas of how to make a simple monaural or 2-channel stereo source into 5 channels are given in Chapter 4.

There are two components to describe spatial sound: spaciousness and envelopment. These terms are often used somewhat interchangeably, but there is a difference. Spaciousness applies to the extent of the space being portrayed, and can be heard over a 2-channel or a 5-channel system. It is controlled principally by the ratio of direct sound to reflections and reverberation. On a 2-channel system, the sound field is usually constrained to being between the loudspeakers, and spaciousness applies to the sense that there is a physical space portrayed between the loudspeakers. The depth dimension is included, but the depth extends only to the area between the loudspeakers.

Envelopment, on the other hand, applies to the sensation of being surrounded by sound, and thus being incorporated into the space of the recording, and it requires a multichannel sound system to reproduce. Two-channel stereo can produce the sensation of looking into a space beyond the loudspeakers; multichannel stereo can produce the sensation of being there.

Lessons from Concert Hall Acoustics

A number of factors have been identified in concert hall acoustics that are useful in understanding the reproduction of sound over multichannel systems:

- The amount of reverberation, and its settings such as reverberation time, time delay before the onset of reflections, amount of diffusion, and so forth are very important to the perception of envelopment, which is generally a desirable property of a sound field.
- Early reflections from the front sides centered on ±55° from straight ahead (with a large tolerance) add to auditory source width (ASW) and are heard as desirable.
- All directions are helpful in the production of the feeling of envelopment, so reverberation returns and multichannel ambiences should apply to all of the channels, with uncorrelated sources for each channel, although some work shows that a difference at the two ears is preferred for contributions to envelopment. Thus the most important 2 channels for reverberation are LS and RS, the next most important are L and R, and C has little importance.
- Research has shown that 5 channels of direct sound are the minimum needed to produce the feeling of envelopment in a diffuse sound field, but the angles for such a feeling do not correspond to the normal setup. They are ±36°, ±108°, and +180° referenced to center at 0°. Of these, the ±36° corresponds perceptually with ±30° and ±108° with ±110°, but there is no back channel in the standard setup. Thus the sensation of complete diffuse envelopment with the standard 5.1 channel setup is problematic.
- Dipole surrounds are useful to improve the sensation of envelopment in reproduction, and are especially suitable for direct/ambient style recording/reproduction.
- Person-to-person differences in sound field preferences are strong. Separable effects include listening level, the initial time delay gap between the direct sound and the first reflection, the subsequent reverberation time, and the difference in the sound field at the two ears.

Rendering 5 Channels over 2: Mixdown

Many practical setups are not able to make use of 5 discrete loudspeaker channels. For instance, computer-based monitoring on the desktop most naturally uses two loudspeakers with one on either side of the video monitor. Surround loudspeakers would get in the way in an office environment. For such systems, it is convenient to produce a sound field with the two loudspeakers that approximates 5-channel sound.

This may be accomplished starting with a process called crosstalk cancellation. A signal can be applied to the right loudspeaker that cancels the sound arriving from the left loudspeaker to the right ear, and vice versa. One step in the electronic processing before crosstalk cancellation are signals that represent ear inputs to just the left ear and right ear. At this point in the system it is possible to synthesize the correct HRTFs for the two ears, theoretically for sound arriving from any direction. For instance, if sound is supposed to come from the far left, applying the correct HRTFs for sound to the left ear (earlier and brighter) compared to sound to the right ear (later and duller), makes the sound appear to the left.

This process is limited. Cancellation requires good subtraction of two sound fields, and subtraction or "nulling" is very sensitive to any errors in level, spectrum, or timing. Thus, such systems normally are very sensitive to listener position; they are said to be "sweet spot dependent." Research aims at reducing this sensitivity by finding out just how much of the HRTFs are audible, and working with the data to reduce this effect. Still, there are complications because the head is usually not fixed in location, but can move around, generating dynamic cues as well as static ones.

One way around sweet spot sensitivity, and eliminating crosstalk cancellation, is headphone listening. The problem with headphone listening is that the world rotates as we move our heads left and right, instead of staying fixed as it should. This has been overcome by using head tracking systems to generate information about the angle of the head compared to a reference 3-D representation, and update the HRTFs on the fly.

A major problem of headphone listening is that we have come to learn the world through our own set of HRTFs, and listening through those made as averages may not work. In particular, this can lead to in-head localization, and front–back confusion. It is thought that by using individualized HRTFs these problems could be overcome. Still, headphone listening is clumsy and uncomfortable for the hours that people put in professionally, so this is not an ultimate solution.

In our lab at USC, in the Integrated Media Systems Center, we have used video-based head tracking to alter dynamically the insertion of a phantom stereo image into the two front channels, thus keeping the image centered despite moving left and right, solving one of the major problems of phantoms. This idea is being extended to crosstalk cancellers, and the generation of sound images outside the area between the loudspeakers, in a way which is not so very sweet spot dependent as other systems. A remarkable finding is in demonstrating this to naïve and to professional listeners: the naïve listener accepts that the phantom center should stay centered as he moves left and right and finds it true, whereas the professional listener, having years of experience in hearing such things, is often disturbed by what they here: they cannot reconcile what they hear with their experience.

Direct mixdown from 5 channels to 2 is performed in simpler equipment, such as digital television receivers equipped to capture 5-channel sound off the air, but play it over two loudspeakers. In such a case, it is desirable to make use of the mixdown features of the transmission system that includes mixdown level parameters for center to the 2 channels, and left surround to left and right surround to right channels. The available settings at the production end are: center at −3, −4.5, and −6 dB into each left and right; left surround at −3, −6 dB and off into left, and vice versa into right. For center insertion, −3 dB corresponds to what is needed for a power addition, and it applies completely in the reverberant field of a room; −6 dB on the other hand is a phasor addition that might occur if one were completely dominated by direct sound. It must be said that −4.5 dB is the "right" value, ±1.5 dB!

Auralization and Auditory Virtual Reality

Auralization is a process of developing the sounds of rooms to be played back over a sound system, so that expensive architecture does not need to be built before hearing potential problems in room acoustics. A computer model of a sound source interacts with a computer model of a room, and then is played, usually over a 2-channel system with crosstalk cancellation described above. In the future, auralization systems may include 5.1-channel reproduction, as in some ways that could lighten the burden on the computer, since the multichannel system renders sound spatially in a more complete way than does a 2-channel system.

The process can also be carried out successfully using scale models, usually about 1:10 scale, using the correct loudspeaker, atmosphere, and scaled miniature head. The frequency translates by the scale factor. Recordings can be made at the resulting ultrasonic frequency and slowed down by 10:1 for playback over headphones.

Auditory Virtual Reality applies to systems that attempt to render the sound of a space wherein the user can move, albeit in a limited way, around the space in a realistic manner. Auralization techniques apply, along with head tracking, to produce a complete auditory experience. Such systems often are biased towards the visual experience, due to projection requirements all around, with little or no space for loudspeakers that do not interrupt the visual experience. In these cases, multichannel sound from the corners of a cube is sometimes used, although this method is by no means psychoacoustically correct. In theory, it would be possible, with loudspeakers in the corners and applying crosstalk cancellation and HRTF synthesis customized to an individual to get a complete experience of "being there." On the other hand, English mathematician Michael Gerzon has said that it would require one million channels to be able to move around a receiving space and have the sound identical to that moving around a sending space.

Beyond 5.1

The 5.1-channel systems are about three decades old in specialized theater usage, over two decades old in broad application to films in theaters and just a little later in homes, and expanding in broadcasting, supported by millions of home installations that include center and surround loudspeakers that came about due to the home theater phenomena.

Perceptually we know that everyone equipped with normal hearing can hear the difference between mono and stereo, and it is a large difference. Under correct conditions, but much less studied, is the fact that virtually everyone can hear the difference between 2-channel stereo and 5.1-channel sound as a significant improvement. The end of this process is not in sight: the number of channels versus improvement to perception of the space dimension is expected to grow and then to find an asymptote, wherein further improvement comes at larger and larger expense. We cannot be seen to be approaching this asymptote yet with 5.1 channels, but we routinely compare 1, 2, 5.1, and 10.2 channel systems in our laboratory, and while most listeners find the largest difference to be between 2 and 5.1, nonetheless all can hear the difference between 5.1 and 10.2, so we think that the limit of perception has not yet been reached. NHK in Japan has experimented with a 22.2 channel system as well.

One way to look at the next major step beyond 5.1 is that it should be a significant improvement psychoacoustically along the way towards ultimate transparency. Looking at what the 5.1-channel system does well, and what it does poorly, indicates how to deploy additional channels:

- Wide-front channels to reproduce the direction, level, and timing of early reflections in good sounding rooms are useful. An unexpected benefit of the use of such channels was finding that sources panned from front left, through left wide, to left surround worked well with its intermediate phantoms to create the sound of speech moving smoothly around one, something that a 5.1-channel system cannot do.

- A center back channel is useful to "fill in the gap" between surrounds at $\pm 110^{\circ}$, and permits rear imaging as well as improvements to envelopment.
- After the above 3 channels have been added to the standard 5, the horizontal plane should probably be broken in favor of the height sensation, which has been missing from stereo and most multichannel systems. Two channels widely spaced in front of and above the horizontal plane are useful.
- The 0.1-channel may be extended to reproduction from many locations with separate subwoofers to smooth the response. In a large cinema, I conducted a test of six large subs centered under the screen versus four in the corners of the theater, and the four outperformed the six in smoothness of response, and response extension towards both low and high frequencies. One electrical channel on the medium drove all the subwoofers, but the response was improved by using multiple ones spaced apart.

Adding these together makes a 10.2-channel system the logical next candidate. Blauert says in his book *Spatial Hearing* that 30 channels may be needed for a fixed listening position to give the impression that one is really there, so 5.1 is a huge step on the road above 2 channels, and 10.2 is still a large step along the same road, to auditory nirvana.

The great debate on any new digital media boils down to the size of the bit bucket and the rate at which information can be taken out of the bucket. We have seen the capabilities of emerging media in Chapter 5. Sample rate, word length, and number of audio channels, plus any compression in use, all affect the information rate. Among these, sample rate and word length have been made flexible in new media so that producers can push the limits to ridiculous levels: if 96 kHz is better than 48, then isn't 192 even better? The dynamic range of 24 bits is 141 dB, which more than covers from the threshold of hearing to the loudest sound found in a survey of up-close, live sound experiences, and at OHSA standards for a one-time instantaneous noise exposure for causing hearing damage.

However, on most emerging media the number of channels is fixed at 5.1 because that is the number that forms the marketplace today. Still, there is upwards pressure on the number of channels already evident. For instance, Dolby Surround EX and DTS ES provide a 6.1-channel approach so that surround sound, ordinarily restricted to the sides perceptually in theaters, can seem to come from behind as well as the sides. 10.2-channel demonstrations have been held, and have received high praise from expert listeners. The future grows in favor of a larger number of channels, and the techniques learned to render 5.1-channel sound over 2-channel systems can be brought to play for 10.2-channel sound delivered over 5.1-channel systems.

The future is expected to be more flexible than the past. When the CD was introduced, standards had to be fixed since there was no conception of computers capable of carrying information about the program along with it, and then combining that information with information about the sound system, for optimum reproduction. One idea that may gain favor in time as the capacity of media grows is to transmit a fairly high number of channels, such as 10.2, and along with it metadata about its optimum presentation over a variety of systems, that could range from a home theater, to car stereo, to headphone listening. Even two different presentations of 10.2 could be envisaged: why not have both the best-seat-in-the-house and the middle-of-the-band perspective available from one program file, merely by changing the metadata? We have done this experimentally with a recording of Messiah, producing both the best seat approach for primary listening and the middle of the chorus approach for participation by singing along.

Addendum: The Use of Surrounds in *Saving Private Ryan*

Gary Rydstrom

Since we hear all around us, while seeing only to the front, sounds have long been used to alert us to danger. In Steven Spielberg's *Saving Private Ryan*, the battle scenes are shot from the shaky, glancing, and claustrophobic point of view of a soldier on the ground. There are no sweeping vistas, only the chaos of fighting as it is experienced. The sound for this movie, therefore, had to set the full stage of battle, while putting us squarely in the middle of it. I can honestly say that this film could not have been made in the same way if it were not for the possibilities of theatrical surround sound. If sound could not have expressed the scale, orientation, and emotion of a soldier's experience, the camera would have had to show more. Yet it is a point of the movie to show how disorienting the visual experience was. Sound becomes a key storyteller.

The sound memories of veterans are very vivid. We started our work at Skywalker Sound on *Saving Private Ryan* by recording a vast array of World War II era weapons and vehicles. In order to honor the experiences of the men who fought at Omaha beach and beyond, we wanted to be as accurate as possible. I heard stories such as how the German MG42 machine gun was instantly identifiable by its rapid rate of fire (1,100 rounds a minute, compared to 500 for comparable Allied guns); the soldiers called the MG42 "Hitler's Zipper" in reference to the frightening sound it made as the shots blurred into a steady "zuzz." The American M1 rifle shot eight rounds and then made a unique "ping" as it ejected its empty clip. Bullets made a whiney buzz as they passed close by. The German tanks had no ball bearings and squealed like metal monsters. These and many other sound details make up the aural memories of the soldiers.

Our task was to build the isolated recordings of guns, bullets, artillery, boats, tanks, explosions, and debris into a full out war. Since it isn't handy or wise to record a real war from the center of it, the orchestrated cacophony of war had to be built piece by piece. But, of course, this is what gives us control of a sound track, the careful choosing of what sounds make up the whole. Even within the chaos of a war movie, I believe that articulating the sound effects is vital; too often loudness and density in a track obscure any concept of what is going on. We paid attention to the relative frequencies of effects, and their rhythmic, sequential placement, but also we planned how to use the 6 channels of our mix to spatially separate and further articulate our sounds.

There are many reasons why a sound is placed spatially. Obviously, if a sound source is on screen we pan it to the appropriate speaker, but the vast majority of the sounds in the *Saving Private Ryan* battles are unseen. This gave us a frightening freedom in building the war around us. To handle this freedom, we kept in mind several purposes for the surround channels.

Overcoming the Masking Effect

It is often amazing, even to sound mixers and editors, how sounds can be altered by context. As frequencies compete when sounds are added together, the characters of the sounds change, sometimes disappearing into each other. Multiple channels can be used to separate sound effects that would mask each other if sharing a channel. As an example, if I have a low frequency explosion in the front channels, I would at that time make sure the background artillery or the rumble of a ship engine is in the rear channels. If there is sand debris from the explosion in the left channel, I would choose to have bullet passbys—which are in the same frequency range—in the right or rear channels. The separate boom channel helps in this way, too, by taking some of the load of playing the very low frequencies away from the main channels. This "orchestration" of sounds is very important. Speaker separation helps overcome the muddy results of masking effects.

Orientation

When we land on Omaha beach, the action is staged such that the German defenses tend to be to our right and behind us, while the incoming American troops tend to the front and left. Sound helps us orient to this by placing outgoing artillery in the back and right channels, the resulting explosions in the front, the American guns to the left, and the German guns behind us. In this way, the sound track gives us some bearing while the camera shakes along the ground, seeing little beyond our immediate surroundings. Split surround channels are very helpful in these circumstances. Originally, Dolby Surround was meant to envelop us in one large, mono, ambient field. Now that surround channels are more full range, and differentiated left and right, the sound can orient us more accurately. Full-range sounds such as explosions and gunfire can be placed more usefully around us; in a war movie these big sounds do constitute "ambience." In film editing there is a concept of "stage line," which is an imaginary line the camera usually does not cross in order to help our geographical understanding of a scene; in an interesting way, the 6-channel system allows the sound to follow this "stage line" principle more successfully.

Contrast

The more tools we have to shift a sound track, the more dramatic we can be. In action scenes, particularly, we can use contrasting frequencies, volumes, and spatial placement to maintain a scene's impact. "Dynamics" of all types are central to much film sound. Spielberg designed some ingenious moments of contrast into the Omaha beach battle: first, the camera dives below water several times giving us brief but dramatic breaks from the sounds of all-out battle; second, Miller is shell-shocked into temporarily losing his hearing. In both cases, the track not only gets quieter, and subjectively stranger, but the multichannel sound image narrows to a simpler stereo. The contrast of all these elements draws us into a different point of view for a time, before we are blasted back to all 6 channels of reality.

Movement of Sounds

The most dramatic use of surrounds in movies involves moving sounds through speakers. This amounts to the "cool" reason for spatially placing sounds. In a film sound track, it is the change in sound we notice most, so the change of location through multiple speakers gets our attention. In *Saving Private Ryan* bullet passbys are moved through the space very often, usually from back to front preceding a bullet impact on screen. Artillery shells whiz by us before exploding. Gun echoes slap off one side of the theater before slapping off the other. The movement of these sounds gives us the impression of being in the action, having the closest possible proximity to the horrors, and of being as unsafe as the men onscreen. Sound, in this way, can take us through the classical proscenium and put us in the story, in a way that smaller formats like television cannot. Moving sounds do not just support things passing by us, but they also support a moving point of view: as the camera swings around to look at the encircling carnage, the sounds of screaming men and guns swirl around in kind. This is a very effective way to draw an audience into the minds of the characters, and into the drama onscreen. The current 6-channel sound systems allows us to treat the theater space as a circle, and the resulting sense of being "surrounded" is literally and figuratively beneficial to *Saving Private Ryan*.

The Limitations of Surrounds

There are some important limitations to keep in mind when mixing in a surround format. First, surround channels are not as full range as the front channels; sounds that pan through them lose some quality, and won't have equal weight in the back. Second, the left and right surround channels are made up of long arrays of speakers, so sounds do not localize to a point as they can up front; there is still a vestige of the old ambient purpose of surrounds. Third, there can be a fine line between encompassing an audience and distracting them; mixing requires making choices about using the surrounds in ways that don't artificially pull the audience out of the action. This is a subjective choice, but it becomes obvious with experience that there are times to surround the audience with sounds, and there are times not to. For example, as we cut to close ups of an American sniper and a German sniper standing off in a courtyard, the sound track narrows to the rhythmic tapping of rain drips on the respective guns, using the surrounds less and "focusing" us on the immediate, intimate actions of the snipers. Just as focus, framing, depth of field, and camera movement guide what we are meant to see, similar techniques—including the use or non-use of surrounds—are available for sound.

The great power of sound in film is its emotional power. Like the sense of smell, sound can conjure up memories, but even more importantly it can conjure up feelings. In *Saving Private Ryan* the audience is asked to share feelings of anxiety, horror, fear, courage, confusion, friendship, loss, safety, and danger. Sometimes the visual violence is so much that we want to close our eyes, but—thankfully for my job—we can't close our ears. Our aural processing of the world, and of movies, is constant, and therefore oftentimes unappreciated. The encompassing sound possible in today's film theaters is so important to *Saving Private Ryan* simply because it simulates how those soldiers heard their world. When we get into their heads, we can better share their experience.

Surround sound is not just useful to action and war movies. As with every element of filmmaking, sound should be used to tell a story. The unique ability of film sound today is in telling the story of what is happening off-screen. There can be amazing power in that. In many ways, the advance of home theater systems has greatly increased the awareness of sound, especially surround sound. But there is a danger in not integrating the use of 6-channel sound with a film's story and purposes, and there is more to this latest sound technology than making good demo discs. Yet when there is a combination of good filmmaking and thoughtful sound work, we can be drawn into an experience like no other.

Appendix 1 Sample Rate

Of the three items contending for space on a digital medium or for the rate of delivery of bits over an interface, perhaps none is so contentious as sample rate, which sets the audio frequency range or bandwidth of a system. The other two contenders are word length (we're assuming linear PCM coding) that sets the dynamic range, and the number of audio channels. Since sample rate and word length contend for the precious resource, bits, it is worthwhile understanding thoroughly the requirements of sample rate and word length for high audible quality, so that choices made for new media reflect the best possible use of resources. Word length is covered in Appendix 2.

Some media such as Digital Versatile Disc Audio (DVD-A) require the producer of the disc to make a decision on the sample rate, word length, and number of channels. Sample rates range from 44.1 to 192 kHz, word lengths from 16 to 24 bits, and the number of channels from 1 to 6, although some combinations of these are ruled out as they would exceed the maximum bit rate off the medium. Furthermore, the producer may also make a per channel assignment, such as grouping the channels into front and surrounds and using a lower sample rate in the surrounds than the fronts. How is a producer to decide what to do?

The arguments made about sample rate from AES papers and preprints, from experience co-chairing the AES Task Force on High-Capacity Audio and IAMM, and from some of those for whom the topic is vital to their work have been collected. First, the debate should be prefaced with the following underlying thesis. The work done 50 years ago and more by such titans as Shannon, Nyquist, and independently and contemporaneously in Russia in producing the sampling theorem and communications theory is correct.

Debates occur about this on-line all the time between practitioners and those who are more theory oriented, with the practitioners asking why "chopping the signal up" can possibly be a good thing to do, and theoretical guys answering with "because it works" and then proceeding to attempt to prove it to the practitioners. For example, "how can only just over two samples per cycle represent the complex waveform of high-frequency sound?" Answer: "because that's all that's needed to represent the fundamental frequency of a high-frequency sound, and the harmonics of a high-frequency sound that make it into a complex waveform are ultrasonic, and thus inaudible, and do not need to be reproduced." When it comes to the question "then why do I hear this?" from the practitioners, that's when things get really interesting: are they just hearing things that aren't there, is what they are telling us is that the equipment doesn't work according to the theory, or is there something wrong with the theory? We come down of the side of the sampling theorem being proved, but there are still a lot of interesting topics to take up in applying the theory to audio.

Here are the arguments made that could affect our choice of sample rate:

1. That the high-frequency bandwidth of human hearing is not adequately known because it hasn't been studied, or we've only just scratched the surface of studying it.

It is true that clinical audiometers used on hundreds of thousands of people are only calibrated to 8kHz typically, due to the very great difficulties encountered in being sure of data at higher frequencies. Yet there are specialized high-frequency audiometer experiments reported in the literature, and there are even industrial hygiene noise standards that set maximum levels beyond 20kHz. How could a scientific basis be found for government regulations beyond audible limits? Is there another effect than direct audition of "ultrasonic" sound?

There is work that shows some young people can hear sine waves in air out to 24kHz, if the level is sufficiently high. From peer-reviewed scientific journals: "the upper range of human air conduction hearing is believed to be no higher than about 24,000Hz." On the other hand, sound conduction experiments with vibration transducers directly in contact with the mastoid bone behind the ear show perception out to "at least as high as 90,000Hz," but the sound that is heard is pitched by listeners in the 8–16kHz range, and is thus likely to be the result of distortion processes rather than direct perception. Certain regulatory bodies set maximum sound pressure level limits for ultrasonic sound since many machining processes used in industry produce large sound pressure levels beyond 20kHz, but the level limits set by authorities are over 100dB SPL, and the perception of such high levels probably has more to do with subharmonic and difference-tone intermodulation processes in the mechanical machining process dumping energy down below 20 kHz from distortion of higher-frequency components than by direct perception. While we usually think of the simplest kind of distortion as generating harmonics of a stimulus sine wave at 2 ×, 3 ×, etc. the fundamental frequency, mechanical systems in particular can generate subharmonics at 1/2 ×, 1/3 × the "fundamental" frequency as well. Some compression drivers do this when they are driven hard, for instance. Difference-tone intermodulation distortion produces one of its products at f_2–f_1, and despite f_2 and f_1 both being ultrasonic, the difference tone can be within the audio band, and heard.

So the 20 kHz limit generally cited does come in for question: while it may be a decent average for young people with normal hearing, it doesn't cover all of the population all of the time. As a high-frequency limit, on the other hand, 24 kHz does have good experimental evidence from multiple sources. If the difference between 20 and 24 kHz is perceptible, then we probably should be able to find listeners who can discriminate between 44.1 and 48 kHz sampled digital audio systems with their respective 22.05 and 24 kHz maximum bandwidths, for instance, although the difference is not even as wide as one critical band of hearing, so the ability to discriminate these two sample rates on the basis of bandwidth is probably limited to a few young people.

Interestingly, on the basis of this and other evidence and discussions, the AES Task Force on High-Capacity Audio several years ago supported sample rates of 60 or 64 kHz, as covering all listeners to sound in air, and providing a reasonably wide transition band between the in-band frequency response and out-of-band attenuation for the required anti-aliasing filter. After a great deal of work had already gone on, it was found that the earliest AES committee on sample rate in the 1970s had also come up with 60 kHz sampling, but that had proved impractical for the technology of the time, and 44.1 and 48 kHz sample rates came about due to considerations that included being "practical" rather than being focused on covering the limits of the most extreme listeners. This is an example of the slippery slope theory of compromise: without covering all of the people all of the time, standards come to be seen as compromised in the fullness of time.

2. The anti-aliasing and reconstruction filters for the ADC and DAC, respectively that are required to prevent aliasing and reconstruction artifacts in sampled audio have themselves got possible audible artifacts. These include potentially: (1) steep sloped filters near the passband having undesired amplitude and phase effects, (2) passband frequency response ripple may be audible, (3) "pre-echo" due to the phase response of a linear phase filter could be audible.

The 24kHz answer deals with the question of steepness of the amplitude response of the filter: if the frequency of the start of rolloff is above audibility, the steepness of the amplitude slope does not matter. The corollary of having a steep amplitude slope though is having phase shift in the frequencies near the frequency limit, called the band edge in filter design. Perhaps the steep slope is audible through its effects on phase? A commonly used alternative to describing the effects of phase is to define the problem in terms of time, called group delay.

Fortunately, there is good experimental data on the audible effects of group delay. Doug Preis, a professor at Tufts University, in particular has studied these effects by reviewing the psychoacoustic literature thoroughly and has published extensively about it. Group delay around 0.25ms is audible in the most sensitive experiments in the mid-range (in headphones, which are shown to be more sensitive than loudspeaker listening), but the limit of having any audible effect goes up at higher and at lower frequencies. The limit is about 2ms at 8kHz, and rises above there.

By comparing the properties of the anti-aliasing and reconstruction filters with known psychoacoustic limits it is found that the effects of the filters are orders of magnitude below the threshold of audibility, for all the common sample frequencies and examples of filters in use. For example, I measured a high-quality group-delay-compensated analog "brick wall" anti-aliasing and reconstruction filter set. The group delay is orders of magnitude less than 2ms at 8kHz, and in fact does not become significant until the amplitude attenuation reaches more than 50dB. Since this amount of attenuation at these frequencies puts the resulting sound at relatively low levels, this is below the threshold of even our most sensitive group of listeners, so the group delay is irrelevant.

What we are discussing here is the group delay of the various frequency components within one audio channel. Shifts occurring between channels are quite another matter. In a head-tracking system operating in the Immersive Sound Lab of the Integrated Media Systems Center at USC, an operator sitting in front of a computer monitor is tracked by video, so that the timing of the insertion of center channel content into the two loudspeakers, left and right of the monitor, can be adjusted as the listener moves around, keeping the phantom image uncannily centered as the listener moves left and right. Part of the procedure in setting this up is to "find the center." Under these circumstances, dominated by direct sound, it is easy to hear a one sample offset between the channels as an image shift. One sample at 48kHz is 20.833μs. It is known that one "just noticeable difference" (jnd, in perceptual jargon) in image shift between left and right ear inputs is 10μs, so the 20μs finding is not surprising. It is also amazing to think that perception works down in time frames that we usually think of as associated with electronics, not with the much more sluggish transmission due to nerve firings, but what the brain is doing is to compare the two inputs, something that it can do with great precision even at the much slower data rate of neuron firing. This is what gives rise to the sensitivity of about 1° that listeners show in front of them for horizontal angular differences, called the minimum audible angle. So here is one finding: systems cannot have even so much as a one sample offset in time between any channel combination, for perceptual reasons (as well as other reasons such as problems with summing).

The difference between intra-channel and inter-channel group delay is probably all too often the source of confusion among people discussing this topic: in the first case 10μs is significant, in the other it takes thousands of microseconds (namely, milliseconds) to become significant. Also, it is not known what equipment meets this requirement. Some early digital audio equipment sampled alternate channels left-right-left-right..., putting in a one sample offset between channels that may or may not have been made up in the corresponding DAC. While most pro audio gear has got this problem licked, there are a large number of other designs that could suffer from such problems, particularly in a computer audio environment where the need for sample level synchronization may not yet be widely understood. And it's a complication in pro digital audio consoles, where every path has to have the same delay as every other, despite having more signal processing in one path than another. For instance, feeding a signal out an aux send DAC for processing in an external analog box, then back in through an aux return ADC, results in some delay. If this were done on say, left and right, but not on center, the position of the phantom images at half left and half right would suffer. In this case, since there is more delay in left and right than in center, the phantoms would move toward the earlier arriving center channel sound.

Another factor might be pre-echo in the required filters. Straight multibit conversion without oversampling makes use of an analog "brick wall" filter for aliasing, but these filters are fairly expensive. It is simpler to use a lower slope analog filter along with a digital filter in an oversampling converter, and these types account for most of the ones on the market today. By oversampling, the analog filter requirements are reduced, and the bulk of the work is taken over by a digital decimation filter. Such digital filters generally have the property that they show "pre-ring," that is, ringing before the main transient of a rapidly rising waveform. Group-delay compensated ("linear phase") analog filters do this as well. Although at first glance this looks like something comes out of the filter (the pre-ring) before something else goes in (the transient), in fact there is an overall delay that makes the filter "causal," that is, what is simply going on is that the high-level transient is more delayed in going through the filter than the content of the pre-ring.

One corollary to the pre-ring in a digital filter is ripple in the frequency response of the passband of the filter. Specification for passband response ripple today in digital filters is down to less than ±0.01 dB, a value that is indistinguishable from flat amplitude response to listeners, but the presence of the ripple indicates that there will be pre-ring in the filter. Since the pre-ring comes out earlier than the main transient, do we hear that?

This problem has been studied extensively in the context of low-bit-rate coders. Perceptual coders work in part by changing the signal from the time domain into the frequency domain by transform, and sending successive frequency response spectra in frames of time, rather than sending the level versus time, as in PCM. The idea is to throw away bits by coding across frequency in a way that accounts for frequency masking, loud sounds covering up soft ones nearby the loud sound in frequency. Since the coding is done in frames of time, and a transient can occur at any time, noise can be left exposed before a transient, potentially heard as small bursts of noise on leading edges that leads to a "blurring effect" on transients. For this reason, a lot of work has gone into finding what the effects are of noise heard just before louder sounds.

Pre-masking is a psychoacoustic effect wherein a loud sound covers up a soft sound that occurs before it. I still remember where I learned this, at a lecture from a psychoacoustician at the MidWest Acoustics Conference many years ago. It seems impossible to anyone with some science background: how can time run backwards? Of course, in fact time is inviolate, and what happens is that the loud sound is perceived by the brain more quickly than the soft one before it, and thus masks it. Armed with data on pre-masking, we can look at the pre-ring effect in filters, and what we find once again is that the pre-ring in practical digital filters in use are orders of magnitude below any perceptual effect. For instance, one data point is that a sound will be masked if it is below –40 dB re the masker level if it occurs 10 ms before the masker.

3. That high-frequency waveshape matters. Since the bandwidth of a digital audio system is at maximum one-half of the sampling frequency, high-frequency square waves, for instance, are turned into sine waves by the sequential anti-aliasing, sampling, and

reconstruction processes. Can this be good? Does the ear recognize waveshape at high frequencies? According to one prominent audio equipment designer it does, and he claims to prove this by making an experiment with a generator, changing the waveform from sine to square or others and hearing a difference. One problem with this experiment is that the rest of the equipment downstream from the generator may be affected in ways that produce effects in the audio band. That is because strong ultrasonic content may cause slewing distortion in power amps, for instance, intermodulating and producing output that is in the audio band. Otherwise, no psychoacoustic evidence has been found to suggest that the ear hears waveshape at high frequencies, since by definition a shape other than a sine wave at the highest audio frequencies contains its harmonics at frequencies that are ultrasonic, and inaudible.

4. That the time-axis resolution is affected by the sample rate. While it is true that a higher sample rate will represent a "sharper," that is, shorter, signal, can this be extended to finding greater precision in "keeping the beat?" While the center-to-center spacing of a beat is not affected by sample rate, the shorter signal might result in better "time-axis resolution." Experiments purporting to show this do not show error bands around data when comparing 48 and 96 kHz sampled systems, and show only about a 5% difference between the two systems. If error bars were included, in all likelihood the two results would overlap, and show there is no difference when viewed in the light of correct interpretation of experimental results.

A related question is that in situations with live musicians, overdub sync is twice as good in a system sampled at twice the rate. While this is true the question is how close are we to perceptual limits for one sample rate versus another. The classic work *Music, Sound, and Sensation* deals with this in a chapter called "The Concept of Time." The author says "If the sound structure of the music is reduced to the simplest sound units, which could be labeled information quanta, one finds an average of 70 units/s which the peripheral hearing mechanism processes." This gives about a 14 ms perceptual limit, whereas the difference improvement by doubling the sample rate goes from 20 to 10 μs. The perceptual limit, and the improvement that might come about by faster sampling, are about three orders of magnitude apart.

The greater problem than simple "time-axis resolution" is comb filtering due to the multiple paths that the digital audio may have taken in an overdubbing situation, involving different delay times, and the subsequent summation of the multiple paths. The addition of delayed sound to direct sound causes nulls in the frequency response of the sum. This occurs because the delay places a portion of the spectrum exactly out of phase between the direct and delayed sound, resulting in the comb. At 48 kHz sampling, and with a one sample delay for ADC and one for DAC, the overall delay is 41.6 μs. Adding a one sample delayed signal to direct sound in 1:1 proportion yields a response that has a complete null at 12 kHz, which will be quite audible.

A related problem is that digital audio consoles may, or may not, compensate for signal processing time delay. Say that one channel needs to be equalized compared to another. That channel will have more inherent delay to produce the equalization. If the console does not compensate, then subsequently adding the equalized channel to an unequalized channel would result in audible comb filters. Thus good designs start with a fixed delay in each channel, then remove delay equal to the required additional processing when it is added. For the worst case, a "look-ahead" style limiter, the range of this effect could be up to 50 ms (which would also place the audio out of lip sync with picture!). One problem with this idea is that if all of the paths through a console have maximum delay all of the time, then live recording situations can have sync problems, such as feeding the cue sends 50 ms late if it is downstream of the maximum delay limiter, which is clearly out of sync.

5. That operating equalizers or in-band filters near the folding frequency distort the frequency response curve of the equalizer or filter dramatically. In a digital console with a bell-shaped boost equalizer centered at 20 kHz, using sampling at 48 kHz, the effect is dramatic because the equalizer must "go to zero" at 24 kHz, or risk aliasing. Thus the response of the equalizer will be very lop-sided. This effect is well known and documented and can be overcome in several ways. One is to pre-distort the internal, digital representation of the desired response in such as way as to best represent the equivalent analog equalizer. This is difficult though possible. A simpler way is to oversample locally within the equalizer to prevent working close to the folding frequency, say a 2:1 oversampling with the required interpolation and decimation digital filters. These filters will make the total impulse response of the system worse, since they are in cascade with the anti-imaging and reconstruction filters, but we are so many orders of magnitude under audibility that we can afford a fair amount of such filtering in the path.
6. A claim has been heard that sampling at, say, 44.1 kHz puts in an "envelope modulation" due to aliasing distortion. This could occur because the limit on frequency response in the pass band, say –0.2 dB at 20 kHz, and the required essentially zero output, say –100 dB by 22.05 kHz, cannot be adequately accomplished, no filter being steep enough. The result is components just above 22.05 kHz being

converted as just below 22.05 kHz (in effect, reflecting off the folding frequency), aliasing distortion that then intermodulates or beats with the signal itself in the ear resulting in an audible problem. This one seems unlikely in practice because high frequencies are typically not at full level where the very non-linear distortion would be most noticeable, and are transient in nature. Also, no experimental evidence is presented to support the idea. As an old AES preprint pointed out, we could look at the effect of phases of the moon on distortion in power amps, but there are probably a whole lot of other things to consider first.

Higher sampling rates are not a one-way street. That is, all items to consider don't necessarily improve at higher sampling rates. Here is a prime consideration: doubling the sample rate halves the available calculating power of DSP chips. Thus half as much can be done, or twice as much money has to be spent, to achieve a given result.

While doubling the sample rate also doubles the frequency of the first null frequency when additions are performed that are out of sync, increasing the sample rate is not a solution to this problem: paths should be internally consistent for time delay to the sample.

So what is it that people are hearing with higher than standard sample rates?

- Problems in conventional conversion that may be reduced by brute force using higher sample rates. Note that these are actually faults in particular implementations, not theoretical problems or ones present in all implementations.
- *Aliasing*: In making test CDs, we found audible problems in DACs that involve aliasing, where none should be present. This is easy to test for by ear: just run a swept sine wave up to 20 kHz and listen. What you should hear is a monotonically increasing tone, not one that is accompanied by "birdies," which are added frequencies that seem to sweep down and up as the primary tone sweeps up. This problem is divisible in two: at –20 dBFS the swept sine wave demonstrates low-level aliasing; at near 0 dBFS the demonstration is of high-level aliasing. In theory, digital audio systems should not have birdies, but at least some do. The source for the low-level birdies is probably "idle tones" in oversampling converters, a problem to which such types are susceptible. The source for the high-level birdies is probably internal aliasing filter overload, which is occurring just before clip. Of the two, the low-level ones demonstrate a major problem, while the high-level ones are unlikely to be much of an audible problem on program material, coming as they do only very near overload.

- *Code dependent distortion*: By using a high-level infrasonic ramp with a period of 2 minutes along with a low-level audible tone, the sum of which, along with dither, just reaches plus and minus full scale, it is possible to run through all of the available codes and exercise the entire dynamic range of a DAC or ADC. You simply listen to constant tone and note that it doesn't sound distorted or buzzy over the course of 2 minutes; many DACs sound distorted part of the time. This test was originally designed to test conventional multibit converters for stuck bits, but surprisingly it also shows up audible problems in oversampling types of converters that should be independent of the code values being sent to it. These occur for unknown, but likely practical reasons, such as crosstalk from digital clocks within the digital audio equipment.

Conclusion

In looking thoroughly at the sample rate debate in the AES Task Force, we found support for somewhat higher sample rates than the standard ones such as 60 or 64 kHz, but not so high as 88.2 through 192 kHz. Further, the debate was complicated by two camps, one wanting 96 kHz and the other 88.2 kHz sampling, for easy sample rate conversions from the current rates of 48 and 44.1 kHz. Fortunately, the hard lines of the debate have been softened somewhat since the task force was finished with its work by the introduction of lossless coding of high sample rate audio. Lossless packing, rather like ZIP file compression for audio provides a means to "have your cake and eat it too" due to its lossless storage of, for instance, 96 kHz data at a lower equivalent rate. While this helps mitigate the size of storage requirement for bits due to high sample rates on release storage media, it does not have an impact on the choice of sample rate for the professional audio environment before mastering. Considerations in that environment have been given above.

So what's a producer to do? Probably 99% of the people 99% of the time cannot hear a difference with a sample rate higher than 48 kHz. So most projects will never benefit from higher sample rates in a significant way. But that's not 100%. To reach the 100% goal, 24 kHz bandwidth audio, with a transition band to 30 kHz, sampling at 60 kHz is enough; 96 kHz audio is overkill, but lossless packing softens the pain, making it somewhat practical in release formats. Assigning the highest rate to the front channels, and a lower one to the surrounds, is reasonably practical for direct-ambient style presentation, since air attenuation alone in the recording venue doesn't leave much energy out at frequencies high enough to require extreme sample rates for the surrounds. On the other hand, it can be argued that for mixes embedded "in the band," probably the same bandwidth is needed all round.

What's Aliasing?

What the sampling theorem hypothesizes, is that in any system employing sampling, with a clock used to sample the waveform at discrete and uniform intervals in time, the bandwidth of the system is, at maximum, one-half of the sampling frequency. Frequencies over one-half of the sampling frequency "fold" around that frequency. For instance, for a system without an anti-aliasing input filter employing a 48 kHz sample rate, a sine wave input at 23 kHz produces a 23 kHz output, because that's a frequency less than one-half the sampling rate. But an input of 25 kHz produces an output of 23 kHz too! And 26 kHz in yields 22 kHz out, and so forth. The system operates as though those frequencies higher than one-half the sample rate "bounce" off a wall caused at the "folding frequency" at one-half the sample rate. The greater the amount the input frequency is over the folding frequency, the lower the output frequency is. What has happened is that by sampling at less than two samples per cycle (another way of saying that the sampling rate is not twice the frequency of the signal being sampled), the system has confused the two frequencies: it can't distinguish between 23 and 25 kHz inputs.

One-half the sample rate is often called by various potentially confusing names, so the term adopted in AES 17, the measurement standard for digital audio equipment, is "folding frequency" (Fig. A1-1).

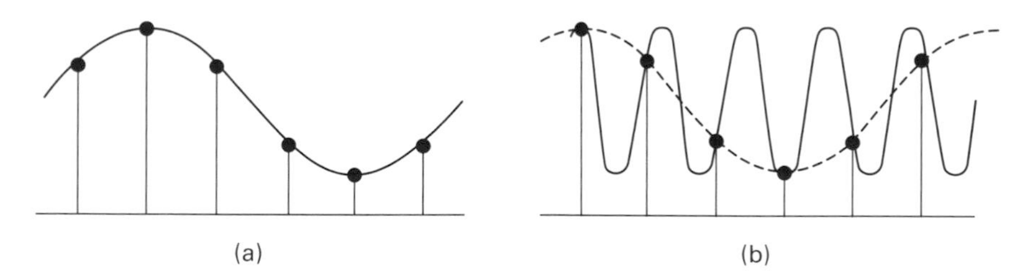

Fig. A1-1 Aliasing is due to sampling fewer than twice per audio cycle and therefore one frequency "looks" to the sampler like another. Another way of looking at this is that the sampling must occur at a frequency of at least twice the highest frequency to be reproduced. Image from John Watkinson, *The Art of Digital Audio*, 3rd ed., Focal Press.

Early digital audio equipment reviews neglected this as a problem. Frequency response measurements on the Sony PCM F1 in one review showed it to be flat to beyond the folding frequency! What was happening was that the frequency response measurement system was inserting a swept sine wave in, and measuring the resulting level out, without either the equipment or the reviewer noticing that the frequencies coming out were not the same ones being put in for the case of those above the folding frequency.

The way around such aliasing is, of course, filtering those frequency components out of the signal before they reach the sampling process. Such filters are called anti-aliasing filters. In a multibit PCM system operating directly at the sample rate, the anti-aliasing filter has to provide flat amplitude response and controlled group delay in the audio pass band, and adequate suppression of frequencies at and above the folding frequency. What constitutes "adequate suppression" is debatable, because it depends on the amplitude and frequency of signals likely to be presented to the input.

Definitions

An anti-aliasing filter is one that has flat response in a defined audio passband, then a transition band where the input signal is rapidly attenuated up to and beyond the folding frequency. The "stop band" of the filter includes the frequency at which the maximum attenuation is first reached as the frequency increases, and all higher frequencies. Practical filters may have less attenuation at some frequencies and more at others, but never have less than a prescribed amount in the stop band. Filter specifications include in-band frequency response, width of the transition region, and out-of-band minimum attenuation.

A reconstruction filter is one used on the output of a DAC to smooth the "steps" in the output. Also called an anti-imaging filter, it eliminates ultrasonic outputs at multiples of the audio band contained in the sample steps.

MultiBit and One-Bit Conversion

In the past, there have been basically two approaches to ADC and DAC design. In the first, sometimes called "Nyquist converters" the input voltage is sampled at the sample rate and converted by comparing the voltage or current in a resistive ladder network to a reference. The resistors in the ladder must be of extremely high precision, since even 16-bit audio needs to be precise to better than 1 part in 65,000. Such designs require complex analog anti-aliasing filters, and are subject to low-level linearity problems where a low-level signal at, say, –90 dB is reproduced at –95 dB.

Delta-sigma ($\Delta\Sigma$) or oversampling converters trade the ultra precision required in level used in a Nyquist converter for ultra precision in time. By trading time for level, such converters are made less expensive, because only a simple anti-aliasing filter is needed, most of the filtering being done in the digital domain, and the high precision resistive ladder is not needed. Unfortunately, conventional $\Delta\Sigma$ converters suffer from several problems, which show up even more when pushing the limits of the envelope to higher sample rates and longer word lengths required by DVD-A. These include limited dynamic range or great expense increases due to the design needs of the accompanying switched-capacitor analog reconstruction filter, susceptibility to idle tones, and high sensitivity to digital noise pickup.

"All 1-bit modulators, regardless of design, have large discrete tones at high frequencies above the audio band (especially clustered around $f_s/2$). This does not cause in-band tones as long as all circuits (downstream) are perfectly linear. Unfortunately, since all circuits are slightly non-linear, these tones fold down into the audio band and cause the now-famous 'idle tones'".[1]

Further work being done today, and reported at conferences as long ago as ISSCC 1998, show designers moving on. "These devices are slowly being taken over by a new generation of multibit (oversampled) converters using a variety of element mismatch shaping techniques such as the one shown in this paper. These new converters offer the hope of high dynamic range (one of the problems of $\Delta\Sigma$ designs) and excellent low-level signal quality (a major problem of Nyquist converters) at low cost."

Converter Tests

The audible tests for DACs and ADCs discussed in the text are available on Vol. 2 of the Hollywood Edge Test Disc Series. To test an ADC, first qualify a DAC as sounding good on these tests, then apply the output of the DAC to the input of an ADC and a second, tested DAC.

The "DAC code check," disc 2 track 19 is an infrasonic ramp with an added low-level 400 Hz "probe tone" to check all code values. The swept sine wave at –20 dBFS is disc 1 track 20. These two tests can be done by ear. The following tests require an oscilloscope. Disc 2 track 15 is a square wave declining from full scale, which checks for overshoot clipping. Disc 2 track 30 is a sine wave that fades up to complete clipping, to check for "wrap around" effects (potential rail-to-rail glitches) in overloaded filters, which are very bad. A current reference for these discs is:

http://www.hollywoodedge.com/product1.aspx?SID=7&Product_ID=95105&Category_ID=12058.

[1] Adams, Robert; Nguyen, Khiem; Sweetland, Karl, "A 112-dB SNR Oversampling DAC with Segmented Noise-Shaped Scrambling," AES Preprint 4774, www.aes.org/e-lib

Appendix 2 Word Length, Also Known as Bit Depth or Resolution

Linear Pulse Code Modulation (LPCM) dominates all other audio coding schemes for high-quality original recording for a number of reasons. First, it is conceptually simple and has had the most research and development compared to other methods of coding. Second, for professionals who manipulate signals, the mathematics of digital signal processing such as equalization is most easily performed on linearly represented signals. Third, it has been shown that the theory of LPCM is just that, linear, that is, distortion free in its theoretical foundation even to well below the smallest signal that can be represented by the least significant bit, when the proper precautions are taken.

As new media come onto the scene, producers may have a choice of word length for representing the audio. Digital Versatile Disc Audio (DVD-A) offers 20- and 24-bit word lengths in addition to 16 bit, and Meridian Lossless Packing allows for one-bit increments from 16 to 24 bits. Once again, the producer is faced with a choice, just as in the choice of sample rate discussed in the previous appendix. While the marketing race may dictate 24-bit audio, reality is, as always, a bit more complicated. Word length is important to professionals because it is one of the three contenders, along with sample rate, and number of audio channels, for space on a medium and transfer rate on and off a medium.

Conversion

Word length in LPCM delivers the dynamic range of the system, just as sample rate delivers the frequency range. It is a familiar story, but there are some nuances that may not be widely known. While the number for dynamic range is roughly 6 dB for each bit, to be precise one has to examine the quantization process in a little more detail. For each short snippet of time, at the sample rate, a device at the heart of digital audio called a quantizer picks the closest "bin" to represent the amplitude of the signal for that sample from one of an ascending series of bins, and assigns a number to it. In LPCM, all the bins have the same "height." There is a likely residual error, because the quantizer cannot represent a number like 1/3 of a bin: either it's in this bin or the next, with nothing in between. Only a few of the samples will probably fall exactly on a bin height; for all other samples there is a "quantization error."

The name given to this error in the past has sometimes been "quantization noise." That's a lousy term because once you've heard it, you'd never call it a mere "noise." A better name is "quantization distortion," which comes closer to offering the correct flavor to this problem—a distortion of a type unique to digital audio. A low-level sine wave just barely taller than one bin crosses over from that bin to the next and gets converted as two bin values, alternating with each other, at the frequency of the sine wave. At the other end of the chain, upon digital-to-analog conversion, reconstruction of the analog output produces a square wave, a terrifically severe distortion of the input sine wave. Even this simplified case doesn't tell the whole story; the sheer nastiness of quantization distortion is underestimated because at least here the sine wave is converted to a square wave, and a sine wave and a square wave of the same frequency are related by the fact that the square wave has added harmonics (1/3 of the level of the fundamental of 3rd harmonic, 1/5 of 5th harmonic, 1/7 of 7th...). Quantization distortion in many cases is worse, as inharmonic frequencies as well as harmonic ones are added to the original through the process of quantization. Also it should be pointed out that once quantization distortion has occurred on the conversion from analog to digital, there is no method to reduce the distortion: the signal has been irrevocably distorted.

Dither to the Rescue

The story on how to overcome this nasty distortion due to quantization is a familiar one to audio professionals: add appropriate dither. Dither is low-level noise added to linearize the process. Dither noise "smears" the bin boundaries, and if the correct amount and type are used, the boundaries are no longer distinguishable in the output, and in theory the system is perfectly linear. Tones below the noise floor that once would not have triggered any conversion because they failed to cross from one bin to the next are now distinguished. This is because, with the added noise, the likelihood that tones lower than one bit cross the threshold from one bin to the next is a function of the signal plus the dither noise, and the noise is agitating the amplitude of the signal from sample to sample up and down. On the average of multiple samples, the tone is converted, and audible, even below the noise floor.

The story of the origin of the technique of adding dither to linearize a system is in Ken Pohlmann's book *Principles of Digital Audio*. World War II bombsights were found to perform better while aloft than on the ground. Engineers worked out that what was going on was that the vibration of flight was averaging out the errors due to stiction and backlash in the gear arrangements in the bombsights. Vibrating motors were added so that the sights would perform better on the ground. Dither is also used today by picture processing programs such as PhotoShop, so that a gradually graded sky does not show visible "bands" once quantized.

The type of dither that is best for audio is widely debated, with some digital audio workstations offering a variety of types selectable by the producer (who sometimes seems in today's world to be cursed with the variety of choices to be made!). One well researched dither noise that overcomes quantization distortion, and also noise modulation (which is a problem with other dither types), while offering the least additional measured noise, is white noise having a triangular probability density function at a peak-to-peak level of ±1 least significant bit. This is a fancy way of saying that for any one sample, the dither will fall in the range of ±1 LSB around zero, and over time the probability for any one sample being at a particular level follows a triangle function. The jargon for this is "TPDF" dither. But there is a noise penalty for dither: it exchanges distortion for noise, where the noise is a more benign "distortion" of the original signal than is the quantization distortion.

So every LPCM digital audio system requires dither, for low distortion. Often in the past this dither has been supplied accidentally, from analog noise, or even room noise, occurring before conversion. As the quality of equipment before the quantizer improves, and rooms get quieter to accommodate digital recording, dither has become the greatest limitation on signal-to-noise ratio of digital audio systems. The amount of noise added for TPDF dither is 4.77 dB, compared to the theoretical noise level of a perfect quantizer without dither.

Dynamic Range

So, to dot all the "i's" and cross all the "t's," the dynamic range of a LPCM digital audio ADC conversion, performed with TPDF white-noise dither and no pre- or de-emphasis, is:

$$(6.02n + 1.76) - (4.77)\,\text{dB} = 6.02n - 3.01\,\text{dB}$$

and approximately

$$6n - 3\,\text{dB}$$

where n is the number of bits in a word.

So this is a pretty easy-to-remember new rule of thumb: just subtract 3 dB from six times the number of bits. (We are neglecting the 0.01 dB error in the mismatch of 1.76 and 4.77 dB, and even the 0.02 times the number of bits error inherent in 6.02 dB, but I once got a job with a test that assumed you would round 6.02 dB to 6 dB if you were a "practical" engineer (at Advent), and note that the sum of these errors is only about 1/2 dB at the 24-bit level, and we're talking about noise here, where 1/2 dB errors approach measurement uncertainty.)

The theoretical dynamic range for 16-, 20-, and 24-bit LPCM systems is given in Table A2-1. Table A2-2 shows the corresponding signal-to-noise ratios, using the Society of Motion Picture and Television Engineers (SMPTE) reference level of −20 dBFS. See the section at the end of this appendix about this reference level in its relationship to analog film and

Table A2-1 LPCM Dynamic Range

Number of bits	***Dynamic range with TPDF, white-noise dither and perfect quantizer (dB)***	***Dynamic range as at left, A weighted (dB)***
16	93.3	96.3
20	117.4	120.4
24	141.5	144.5

Table A2-2 LPCM Signal-to-Noise Ratio, −20 dBFS Reference

Number of bits	***Signal-to-noise ratio with TPDF, white-noise dither and perfect quantizer***	***Signal-to-noise ratio as at left, A weighted***
16	73.3	76.3
20	97.4	100.4
24	121.5	124.5

tape levels. Also note that there are special requirements for multichannel dither, described by Robert A. Wannamaker in "Efficient Generation of Multichannel Dither Signals," *JAES*, Vol. 52, No. 6, 2004 June, pp. 579–586.

For high playback levels, the noise floor of 16-bit digital audio with a TPDF, white-noise floor, intrudes into the most sensitive region of human hearing, between 1 and 5 kHz. Thus action is called for to reduce noise, and there are several ways to do this. But first, let us think about whether we are in fact getting 16-bit performance from today's recordings.

Large film productions are using mostly 16-bit word length for their recordings. Let us just consider the noise implications on one generation of a multiple generation system, the stage of premixes summing to the final mix. Perhaps eight premixes are used, and "Zero Reference" mixes are the order of the day. This means that most of the heavy work has already been done in the pre-mixing, allowing the final mix to be a simple combination of the premixes, summed together 1:1:1.... Now eight noise sources that are from different noise generators add together in a random manner, with double the number of sources resulting in 3 dB worse noise. Therefore, for our eight premix to final mix example, the final mix will be 9 dB noisier than one of the source channels (1:2 channels makes it 3 dB worse, 2:4 another 3 dB, and 4:8 another 3 dB, for a total of 9 dB). Our 76 dB weighted signal-to-noise ratio from Table A2-2 for one source channel has become 67 dB, and the actual performance heard in theaters is 14.5 bits!

Similar things happened in music recording on 24-track analog machines, mixed down to 2-track. Noise of the 24-tracks is summed into the two, albeit rarely at equal gain. Thus the noise performance of the 2-track is probably thoroughly swamped by noise from the 24-track source, so long as the two use similar technology.

These examples of noise summing in both analog and digital recording systems illustrate one principle that must not be overlooked in the ongoing debate about the number of bits of word length to use for a given purpose: professionals need a longer word length than contained on release media, because they use more source channels to sum to fewer output channels, in most cases. Only live classical recording direct to 2-track digital for release does not require any longer word length, and then, if you want to be able to adjust the levels between recording and release, you should have a longer word length too. So there is virtually no professional setting that does not require a longer word length than release media, just to get up to the quality level implied by the number of bits of word length on the release.

Actual Performance

Today's conversion technology has surpassed the 16-bit level. On the other hand, I have measured a "24-bit" equalizer and its dynamic range was 95dB, just 16-bit performance. So what's the difference between "marketing bits" and real ones? The dynamic range determines the real word length, and how close a particular part comes to the range implied by the number of bits, not the top of the data page that screams 24 bit. As an example, here are two of the widest dynamic range converters on the market. One ADC converter chip has 123dB dynamic range, A weighted. A corresponding DAC also has 123dB dynamic range. Both specifications are "typical," not worst case which is some 6dB worse, so these are not a basis for specifying a finished unit made with them. Each of the ADC and DAC actually performs practically at 20.5-bit dynamic range (although it is A weighting that delivers that extra 1/2 bit). For the pair, the noise power adds, and the total will be 120dB dynamic range A weighted, or call it 20-bit performance. By the way, this "matching" of dynamic range between ADC and DAC is not very common.

There are ways to get "beyond the state of the art." By paralleling multiple converters, the signal is the same in all the converters, but the noise is (supposedly) random among them. By summing the outputs of paralleled converters together, the contribution of the signal grows faster (twice as much in-phase signal is 6dB) than the contribution of the noise (twice as much noise is 3dB, so long as the noises being added are uncorrelated), so noise is reduced compared to the signal and dynamic range is increased. Using our example of the premixes in reverse, eight each of the ADCs and DACs, paralleled, can theoretically increase the dynamic range by 9dB, or 1.5 bits. So the 20-bit pair becomes 21.5 bits, or 129dB, if you're willing to put up with the cost and power of 8 converters for each of the ADCs and DACs in parallel.

It is interesting to compare this range with that of some microphones readily available on the market. The Neumann TLM-103 that is a rather inexpensive one for its type has 131dB dynamic range, so surpasses our complicated and thus probably theoretical paralleled converter by several dB. This fact is what has given impetus to the method of using several ADCs to cover various scaled ranges of the dynamic range, and then combine the outputs in DSP codes in a clever way, to reach even more than the dynamic range of this wide-range microphone. This may be accomplished inside the microphone body itself which then puts out AES-3 code directly, or may be an outboard microphone preamplifier. An example of the first of these is the Neumann D-01 that is specified at 130dB dynamic range with the capsule connected, and

140 dB with the input of the ADC shorted, showing that the conversion noise is not contributing to the output significantly. An example of an auto-ranging microphone preamplifier with a clever matching scheme so that there are no glitches is the StageTec TrueMatch model with 153 dB A weighted specified dynamic range, so this goes even beyond the ability of AES 3 to represent.

How Much Performance Is Needed?

This is the great debate, since word length represents dynamic range, and dynamic range varies from source to source, and its desirability from person to person. Louis Fielder of Dolby Labs did important work by measuring peak levels in live performances he attended, and found the highest peaks to be in the range of 135–139 dB SPL. I measured an actor screaming at about 2 ft away to be 135 dB SPL peak, so this is not a situation limited to just live music. Note that these were measured with a peak-responding instrument, and that an ordinary sound level meter would read from a few, to more likely many decibels less, due to their 1/8-second time constant and their average or rms detectors, when they are set to fast.

For sound from film, levels are more contained, and the maximum level is known, because film is dubbed at a calibrated level with known headroom. Together reference level plus headroom produces a capability of 105 dB SPL per channel, and 115 dB for the Low Frequency Enhancement (LFE) channel, which sum, if all the channels are in phase, and below 120 Hz in the range of LFE, to 123 dB SPL. Above 120 Hz, the sum is about 119 dB, near 120 dB that used to be called the Threshold of Pain, although in recent years and with a louder world that designation has been moved up to 140 dB SPL.

So there are pretty big differences between the 119 or 123 dB film maximum, depending on frequency, and the 139 dB live music maximum, but this corresponds to subjective judgment: some live sound is louder than movie sound, and by a lot.

Is the ability to reproduce such loud sounds necessary? Depends on who you are and what you want. Many sound systems of course will not reproduce anything like 139 dB, even at 1 m, much less the 3 m of typical home listening, but I remember well the time I sat in the first row of the very small Preservation Hall in New Orleans about 4 ft in front of the trumpet player—talk about loud and clean. I could only take one number there, and then moved to the back, but it was a glorious experience, and my hearing didn't suffer any permanent shift (but I've only done that once!). I don't know the level because I wasn't measuring it on that occasion, but it was clearly among the loudest to which I've ever voluntarily exposed myself.

Another experience was the Verdi Requiem heard from the 5th row of Boston Symphony Hall with the Boston Symphony and the Tanglewood Festival Chorus. It was so loud I heard a lot of IM distortion that I associate with overloaded analog tape, only in this case it was my hearing overloading!

We can get a clue as to what is needed from the experience of microphone designers, since they are sensible people who face the world with their designs, and they want them to remain undistorted, and to have a low noise floor. A relatively inexpensive design for its class, the Neumann TLM-103 mentioned before, has 131 dB dynamic range, with a maximum sound pressure level (for less than 0.5% THD) of 138 dB, and a noise floor equivalent of 7 dB SPL, A weighted. For the A/D not to add essentially any noise to the noise floor of the microphone, it should have a noise level some 10 dB below the mike noise, meaning the A/D needs 141 dB dynamic range!

Oversampling and Noise Shaping

When the first CD players appeared, Sony took a multibit approach with a 16-bit DAC and an analog reconstruction filter, as described in the appendix on sample rate. Philips, on the other hand, took a different approach, one that traded bandwidth and dynamic range off in a different manner. Audio trades bandwidth and dynamic range all the time, perhaps without its users quite realizing it. For instance, an FM radio microphone has a much wider bandwidth than audio such as ±75 kHz to overcome the fairly noisy FM spectrum in which the system operates (then techniques of pre- and de-emphasis and companding are also applied to reduce noise). Oversampling is a little like FM modulation/demodulation in the sense that oversampling spreads the noise out over a band which is now at a multiplier times the original folding frequency (one-half the sample rate), which means that much of the noise lies above the audio band, and is thus inaudible. The first Philips players used 14-bit converters and oversampling to achieve a 16-bit dynamic range, using this technique, and it is in use more aggressively to this day.

More recently, the idea of psychoacoustic equalization of the dither has risen to prominence. In-band white-noise dither is a particularly bad choice psychoacoustically, although simple electronically, because it weights noise towards higher frequencies when viewed on a perceptual basis, just where we are most sensitive (1–5 kHz). It is better to suppress the noise in this region, and increase it proportionally at higher frequencies to produce the same dither effect. For one such dither spectrum the amount of the "dip" in the noise centered at about 4 kHz is 20 dB, and the increase is nearly 30 dB in the 18–22.05 kHz region. This is enough of an increase that it can often be seen on wide dynamic range peak meters. At first it may seem peculiar to see the noise level on the meters go up, just as the audible noise goes down, but that's what happens. By using F-weighted psychoacoustic noise shaping, a gain of about 3 1/2 bits of audible performance is created within the framework of a particular word length. Thus 16-bit audio, with psychoacoustic noise shaping, can be made to perform like the audible dynamic range of 19 1/2-bit audio.

It may be a bad idea to psychoacoustically noise shape original recordings that are subject to a lot of subsequent processing. If high-frequency boost equalization was involved, the increase in noise due to noise shaping at high frequencies could be revealed audibly, through crossing over from lying underneath the minimum audible sound field, to being above. Better is to use a longer word length in the original recording, then do all subsequent processing, before adding the proper amount of shaped noise to "re-dither" the resulting mix to the capacity of the media.[1]

The Bottom Line

From the discussion above, the following emerges:

- Professionals nearly always need to work at a greater word length in conversion, storage, and mixing than on the release media, for the ability to sum source channels without losing resolution below the requirement implied by the release media word length. Some processes, such as equalization, mixing busses, etc. require even greater resolution than the input–output resolution since they deal with wider dynamic range signals. In particular units that recirculate signals, like reverberation devices, require much longer word lengths than the main system to which they are connected since the recirculation adds noise.
- Wider dynamic range is desirable for professionals (longer word lengths) in original recording A/D conversion and recording than may be necessary on release media, because accidents happen where things are suddenly louder than expected, and it is useful to have extra headroom available to accommodate them. Obviously,

[1]Stanley P. Lipshitz, et. al., "Dithered Noise Shapers and Recursive Digital Filters," *J. Audio Eng. Soc.*, Vol. 52, No. 11, 2004 November, pp. 1124–1141.

once the level has been contained by fader setting downstream of the original source, this motive is less of a consideration.

- The practical limit today on conversion is perhaps 21 bits, except for certain devices that employ multiranging.
- Since the best single ADC today reaches 21 bits, and microphones reach perhaps something more than the same range, 24-bit audio for professional storage seems not unreasonably over specified today.
- Psychoacoustic noise shaping can add more than 3 bits of audible dynamic range, but it must be used sensibly. Dither, on the other hand, is always necessary to avoid quantization distortion. So some projects may benefit from original conversion with TPDF white-noise dither at longer word lengths, and after subsequent signal processing, adding dithered noise shaping when in the final release media format.
- The word length to use for a given project depends on the mixing complexity of the project: longer word lengths are dictated for more complex projects with multiple stages, such as film mixing, compared to simpler ones. The film industry is expected to be changing over from mostly 16-bit systems to mostly 24-bit ones over the next few years.
- The final word length of a release master depends on a number of factors. For a film mix with a known calibrated playback level, there is not much point in going beyond 20-bit audio, since that places the noise floor below audible limits (not to mention background noise levels of rooms). For a music mix with less certain playback conditions, it is hard to foresee a need for greater than 20-bit output, and dithered noise shaping can extend this to nearly an audible dynamic range of 24 bits if needed.

Analog Reference Levels Related to Digital Recording

The SMPTE reference level −20 dBFS is based on the performance of analog magnetic film masters and their headroom, since most movies are stored in analog, and are in the process of being converted to digital video. Unlike the music industry, the film one has kept the same reference level for many years, 185 nWb/m, so that manufacturing improvements to film have been taken mainly as improvements in the amount of headroom available. Today, the headroom to saturation (which gets used during explosions, for instance) is more than 20 dB on the film masters, and some light limiting must be used in the transfer from magnetic film to digital video.

On the other hand, the music industry chased the performance of analog tape upwards over time, raising the reference level from 185 more than 30 years ago, to 250, 320, 355 and even higher reference levels, all in nWb/m, before passing into obsolescence. This outlook takes the headroom as constant, that is, keeping reference level a certain number of decibels below a specific distortion as tape improves, such as −16 or −18 dB re the point of reaching 3% THD. This has the effect of keeping distortion constant and improving the signal-to-noise ratio as time goes by.

I think the difference in these philosophies is due to the differences in the industries. Film and television wanted a constant reference level over time so that things like sound effects libraries had interchangeable levels over a long period of time, with newer recordings simply having more potential headroom than older ones. Also, film practice means that every time you put up a legacy recording you don't have to recalibrate the machine. There is more emphasis placed on interchangeability than in the music industry, where it is not considered a big burden to realign the playback machine each time an older tape is put up.

Appendix 3 Music Mostly Formats

The term Music Mostly* is applied to those formats that, while they incidentally may contain some video or still pictures, devote most of their capacity to audio.

Digital Theater Systems CD

By using fairly light bit-rate reduction, the 1.411 Mbps rate available from the standard CD was made to carry 5.1 channels of 44.1 kHz sampled, up to 24-bit audio by DTS Coherent Acoustics. The signal from such coded discs can be carried by an S/PDIF digital interconnection between CD or Laser Disc players equipped with a digital output and a decoder that may be a stand-alone 5.1-channel decoder, or a decoder built into a receiver or A/V system controller, many of which are available in the marketplace.

DVD-Audio

The audio-mostly member of the DVD family has several extensions and possibilities for audio beyond those offered for Digital Versatile Disc Video (DVD-V). The result of the changes offered for audio is that Digital Versatile Disc Audio (DVD-A) software will not play on existing DVD-V players, unless the DVD-A has been deliberately made with cross DVD compatibility in mind, which would typically be accomplished by including a Dolby Digital bit stream along with the linear PCM (LPCM) high bit rate audio. A Dolby Digital recording would be made in the DVD-V "zone" of a DVD-A. DVD-V players would read that zone, and not see the DVD-A "zone" represented in a new directory called AUDIO_TS. For these reasons, although DVD-A only players may be introduced

*This is a play on words on the long-running New York musical series Mostly Mozart.

for audiophiles, universal DVD players that can handle both video and audio discs will likely be the most popular.

Producers may change sample rate, word length, and number of channels from track to track on DVD-A. Even within one track, sample rate and word length may vary by channel. Tables A3-1 and A3-2 give some samples of how, if all the channels use matched sample rate and word length, the maximum bit rate of 9.6 Mbps is respected, and give the

Table A3-1 LPCM for DVD-A*

Sample rate (kHz)	*Word length (bits)*	*Maximum number of channels*	*LPCM bit rate (Mbps)*	*Playing time DVD-5 (min)*
44.1	16	5.1	4.23	127.7
44.1	20	5.1	5.29	104.2
44.1	24	5.1	6.35	87.9
48.0	16	5.1	4.61	118.2
48.0	20	5.1	5.76	96.3
48.0	24	5.1	6.91	81.2
88.2	16	5.1	8.47	67.1
88.2	24	4	8.47	67.1
96.0	16	5.1	9.22	61.9
96.0	20	5	9.60	59.5
96.0	24	4	9.22	61.9
176.4	16	3	8.47	67.1
176.4	20	2	7.06	79.7
176.4	24	2	8.47	67.1
192.0	16	3	9.22	61.9
192.0	20	2	7.68	73.6
192.0	24	2	9.22	61.9

*LPCM bit rate includes AC-3 448 kb/s stream for backwards compatibility with DVD-V players.

resulting playing times. These two items, bit rate and maximum size of storage, together form the two limitations that must be respected.

"Audio-mostly" means that stills and limited-motion MPEG-2 video content is possible. "Slides" may be organized in two ways: accompanying the sound and controlled by the producer, or separately browsable by the end user. The visual display may include liner notes, score, album title, song titles, discography, and web links. Interactive steering features are included so that menus can be viewed while the sound is ongoing, and playlists can be produced to organize music by theme, for instance.

Table A3-2 MLP for DVD-A*

Sample rate (kHz)	*Word length (bits)*	*Maximum number of channels*	*MLP bit rate peak (Mbps)*	*Playing time DVD-5 (min)*
44.1	16	5.1	1.85	266.6
44.1	20	5.1	2.78	190.1
44.1	24	5.1	3.70	147.7
48.0	16	5.1	2.02	248.9
48.0	20	5.1	3.02	176.6
48.0	24	5.1	4.03	136.9
88.2	16	5.1	3.24	166.2
88.2	20	5.1	5.09	110.7
88.2	24	5.1	6.95	82.9
96.0	16	5.1	3.53	154.2
96.0	20	5.1	5.54	102.3
96.0	24	5.1	7.56	76.6
176.4	16	5.1	4.63	120.7
176.4	20	5.1	8.33	69.8
176.4	24	3	6.88	85.6
192.0	16	5.1	5.04	111.7
192.0	20	4.0	6.91	84.0
192.0	24	3	7.49	78.9
192.0	24	2	4.99	118.3

*Also includes AC-4 bit stream at 448 kbps for backwards compatibility with DVD-V players.

PQ subcodes and ISRC from CD production are utilized. PQ subcodes control the pauses between tracks, and ISRC traces the ownership.

A lossless bit-reduction system is built into the specification and required of all players, affording greater playing time, a larger number of channels at a given sample rate and word length, or combinations of these. Meridian Lossless Packing (MLP) is the packing scheme, and its use is optional. Tables A3-1 and A3-2 show the variations of audio capabilities without and with the use of MLP. Note that the line items in the table emphasize the limiting case for each combination of sample rate and word length, and that programs with fewer channels will generally have longer playing times. In a video "zone" of DVD-A, the other formats Dolby Digital, MPEG Layer 2, and DTS are optional, although Dolby Digital or PCM is mandatory for audio content that has associated full-motion video.

DVD-A has two separate sets of specialized downmixing features, going well beyond the capabilities discussed under metadata earlier. The basic specification allows track-by-track mixdown by coefficient table (gain settings) to control the balance among the source channels in the mix-down and to prevent the summation of the channels from exceeding the output capacity of DACs. As many as 16 tables may be defined for each Audio Title Set, and each track can be assigned to a table. Gain coefficients range from 0 to –60 dB. The feature is called SMART (system-managed audio resource technique). MLP adds dynamic mixdown capability, that is, the ability to specify the gain mixdown coefficients on an ongoing basis during a cut.

MLP extends these choices in several ways, when it is used, which may be on a track-by-track basis on a given disc. With MLP, the word length is adjustable in 1-bit increments from 16 to 24 bits. More extensive mix-down capabilities from multichannel to 2 channel are available than were originally specified in the DVD-A system. MLP adopts a variable bit rate scheme on the medium, for best efficiency in data rates and playing times, and a fixed bit rate at interfaces, for simplicity. The maximum rate is 9.6 Mbps, which is the interface rate all the time, and the peak rate recorded on disc. Coding efficiency gains are shown in Tables A3-1 and A3-2. An example of the use of MLP is that a 96-kHz 24-bit 6-channel recording can be recorded with up to 80 minutes of recording time.

The first data column in Table A3-3, labeled "Minimum gain at peak bit rate," is the reason that MLP was chosen over other lossless schemes. This peak issue was a key factor in selecting MLP instead of the others because while the other lossless coders could show they extended the playing time to a certain amount, there were instances where the peak data rate from the coders still violated the maximum streaming rate of the disc. The MLP compression table describes the compression achieved at the peaks of the VBR output on an average over a longer time, with the peaks being the worst case, and hence have the least compression. So, at 96 kHz and 24 bits, 9 bits per channel are saved on a long-term average by this scheme, or a savings of 37.5%. This gain can be traded off among playing time, sample rate, and word length. It should be noted

Table A3-3 Coding Efficiency Gain for MLP in Units of Data Reduction (bits/sample/channel)

Sample rate f_s (kHz)	***Minimum gain at peak bit rate***	***Long-term average gain***
48	4	8
96	8	9
192	10	11

that the bit rates and playing times given in the table are dependent on the exact nature of the program material, and are estimates.

Due to the variable bit rate scheme, a "look ahead" encoder is necessary to optimize how the encoder is assigning its bits. This results in some latency, or time delay, through the encoder.

Ancillary features of MLP include provision for up to 64 channels (in other contexts than DVD-A); flags for speaker identification; flags for hierarchical systems such as mid-side, Ambisonic B-Format, and others; pre-encoding options for psycho-acoustic noise shaping; and choice of real-time, file-to-file, and authoring options. File-to-file encoding can take place faster than real-time speed, so long as there is enough computer power to accomplish it. The bit stream is losslessly cascadable; contains all information for decoding within the bit stream; has internal error protection in addition to that provided by the error correction mechanisms of the DVD and recovery from error within 2 ms; cueing within 5 ms (decoder only, not transport); and can be delivered over S/PDIF, AES, Firewire, and other connections, although intellectual property concerns may limit these capabilities. Additional data carried alongside the audio include content provider information, signature fields to authenticate copies, accuracy warranty, and watermarking.

The internal error protection of MLP is a very important feature, since LPCM systems lack this feature. Digital audio workstation (DAWs), DTRS and other multichannel tape machines, CD-R and DAT recorders may or may not change the bits in simple, unity-gain recording and playback. In the past, in order to qualify that a digital audio system is bit transparent has required a specialized test signal carried along through all the stages of processing that the desired signals were going to undergo. One such signal is the Confidence Check test pattern built into Prism Media products. Only through the use of this test signal throughout the production, editing, mastering, and reproduction of a set of test CDs was the process made transparent, and there were many stages that were not transparent until problems were diagnosed and solved. MLP has the capability built-in; if it is decoding, then the bit stream delivered is correct.

The tables show some common examples of the use of DVD-A capacity and transfer rates. Table A3-2 shows the effect of LPCM with MLP, including an accompanying Dolby Digital track for backwards compatibility with DVD-V players. The 5.1-channel capacity is calculated using a channel multiplier of 5.25, which represents a small coding overhead. Dolby Digital is calculated at 448 kbps for 5.1-channel audio, and 192 kb/s for 2 channel, although other options are available. MLP supports word lengths from 16 to 24 bits in 1-bit increments, and tracks

on High-Capacity Audio to the capabilities of a medium like DVD-A. The AES Task Force's findings included a sample rate of 60 kHz, a word length of 20 bits, and as many channels as the capacity of a given medium would allow. With DVD-A and MLP, 10.2 channels of audio could be carried within the 9.6 Mbps rate limit, and the disc would play for 84 minutes, when a 5.1-channel Dolby Digital track was included for backwards compatibility with existing DVD-V and DVD-A players.

The conversion from mono to stereo occurred when there was a medium to carry 2 channels, and the conversion from 2 channels to 4 channels occurred when 2-channel media were pressed into the first multichannel service by way of matrix technology. 5.1-channel revolution is well underway, having started in film, proceeded to packaged media and then digital television, and is now reaching the music business. 5.1-channel sound sells in large numbers because its effects are noticeable to millions of people; those effects include superior sound imaging and envelopment. There is no reason to conclude that the end is in sight. The step from 2 to 5.1 channels is just about as noticeable as the step from 1 to 2 channels, or perhaps just a little less so, and while diminishing returns may someday set into the process of adding more channels, exploration of all of the effects of multichannel sound has barely begun.

Index

0.1-channel, 1, 45, 53–64, 181
 See also Low-Frequency Enhancement (LFE)
2-channel stereo, 28, 108, 109, 134, 139
 with 5.1 channel, recording, 100, 102
 downmix, 161–162
 live sound, 118–119
 LPCM, 168
 panning, 112–113
 phantom image, 184–185
 spaciousness, 187
5.1-channel, 56, 191
 additional systems, 191–193
 capacity, 231
 Digital Television, 57–58
 Dolby Digital, 134
 downmix, 161
 history, 9–10
 localization mechanism, 182–183
 LPCM, 12, 147
 mix, 108, 149
 phantom image, 89
 psychoacoustics, 191–193
 specific microphones
  double M-S, 93–94
  Fukada array, 94
  Holophone Global Sound Microphone system, 94
  Ideal Cardioid Arrangement, 95
  OCT array, 95
  Sanken WMS-5, 95–96
  Scheops KFM-360 Surround Microphone, 96
  SoundField microphones, 96
  Trinnov array, 97
 standardized setup, 36–38
6.1-channel system, 47, 192
10.2-channel, 17, 192

A

Absolute polarity, 129
Academy curve, 5
Acoustic shadow, 41, 42, 179
Acoustic summation, 25–26
Advanced Television Systems Committee (ATSC), 151
 audio on DVD-V, differs, 151–152
 metadata, in Digital Television, 151
AES/ITU recommendation, 46
Aliasing, 209, 211–212
Amplitude-phase matrix, 8, 99, 161, 162
An Introduction to the Psychology of Hearing, 178
Anti-aliasing filter, 63, 145, 203–206, 212–213
Apocalypse Now, 9, 114
Applause, 37
Audible tests, 213
Audio coding, 145–148, 151, 171
 cascading coders, 148–149
Audio on DVD-V, 167–169
 and Digital Television, difference, 151–152
Audio production information exists, 160
Audio-to-video synchronization, 163
Auditory Virtual Reality, 190–191
Auralization, 190–191
Auxiliary sends, 121
 in film mixing, 136
 in spot mikes, 136

B

Baby Boom channel, 9, 54, 58
Backwards masking, 148
Barrier-type techniques, 72
Bass management, 25, 56–57, 58, 59
 full-range monitors, 25, 45
 home theater, 59
 Low-Frequency Enhancement, 58, 62, 180–182
 subwoofer, 38
Bass redirection. *See* Bass management
Batman Returns, 165
Benade, Arthur, 182
Binaural microphone, 85–86
 in multichannel, 90–91
Bit depth. *See* Word length
Bit-rate reduction, 12, 14, 134, 147
Blu-ray discs, 16, 169
Boom sting button, 55
Busses, 112, 121
Butterfly, 109, 137–138

C

Cascading coders, 148–149
Center channel, 161
 2-channel stereo, problems of, 36
 downmixing, 161
 equalizing multichannel, 120
 in films, 118
 music mixing, 139
 square array, 49
 surround channels, 91
 "top and bottom" approach, 42
 upmixing, 103
Center loudspeaker, 40–42, 52, 132
Chrétien, Henri, 5
Cinemascope, 5, 6
Cinerama, 5, 6, 47

Close Encounters of the Third Kind, 8, 9, 54
Close-field monitoring, 50–52
Coding gain, 146, 147
Coincident technique
  crossed Figure-8, 80–82
  M-S stereo, 82–83
  for multichannel, 90
  X-Y stereo, 83
Comb filter, 42, 46, 80, 114, 119, 131, 163, 207
Commentary channel, 153
Computer-based monitoring, 188–189
Concert hall acoustics, 30, 188
"Consumer 5.1" mix, 58
Consumer Electronics Association (CEA), 16, 88
Content Scrambling System (CSS), 232
Contribution coders, 149
Control room monitor loudspeakers, 33
Conventional outboard gear, 125
Crossed Figure-8, 80–82
  advantages over M-S stereo, 82–83
Crosstalk cancellation, 189

D

Data essence, 144
Decay time, 32
Decca tree, 72, 74, 78–79, 94, 99
Delivery formats, 141
  audio coding, 145–148
    cascading coders, 148–149
  audio on DVD-V, 167–169
  audio production information exists, 160
  Blu-ray discs, 169
  digital cinema, 172–176
  digital terrestrial, 169, 171
  Digital Versatile Disc, 165–167
  Dolby surround mode switch, 161
  downloadable internet connections, 171
  downmix options, 161–162
  HD DVD, 169
  level adjustment, 162–163
  lip-sync, 163–164
  media specifics, 164–165
  metadata, 151–152
  multiple streams, 152–153
  new terminology, 144
  postproduction, 134–135
  reel edits, 164
  room type, 160
  sample rate, 149–151
  satellite broadcast, 171
  three level-setting mechanisms
    dialnorm, 154–158
    dynamic range compression, 158–159
    mixlevel, 159–160
    night listening, 159
  video games, 171–172
  word length, 149–151
Delta-sigma, 145, 212
Dialnorm, 154–158
  A weighting, 155
  K weighting, 156
  Leq(A), 155
  LKFS, 156
Dialogue channel, 118
Dialogue normalization. *See* Dialnorm
Dialogue service, 153
Digital audio routing, 122
Digital audio workstations (DAW), 78, 92, 110, 112, 150, 217, 231
Digital Cinema, 17, 172–176
Digital terrestrial, 169, 171
Digital Theater System (DTS), 11, 59
Digital Versatile Disc (DVD), 165–167
  See also DVD-A; DVD-V
Direct/ambient approach, 87–88, 115, 121, 138
  panning, 117
  sound microphone technique, 91–93
Direct radiating low-frequency drivers, 33
Direct radiators, pros and cons, 48
Direct-sound all round approach, 15
  Butterfly, 138
  equalization, 121
Direct Stream Digital (DSD), 232
Directional loudspeakers, 27
Directivity index (DI), 33
  coloration, 34
Distribution coders. *See* Mezzanine coder
Dither, 150, 216–217
Divergence, 113, 114, 115
Dolby A noise reduction, 8
Dolby Digital, 11, 12, 14, 134, 152, 165, 227, 231
Dolby E, 124, 134, 149, 163–164
Dolby SR, 8, 9
Dolby Surround EX, 17, 161, 167, 192, 197
Double-system audio accompanying video, 123–124
  Dolby E, 124
  mezzanine coding, 124
Downloadable internet connections, 171
Downmix options, 85, 161–162
Dummy head. *See* Binaural microphone
DVD-A, 144, 146, 201, 215, 227–232
  for LPCM, 228
  for MLP, 229
DVD-V, 144, 151–152, 167–169, 227
DVD music videos
  for surround mixing, 139–140
Dynamic cues, 180
Dynamic panning, 115, 117
Dynamic range, 139, 149, 150, 155, 192, 218–220
  pads and calculations, 103–105
Dynamic Range Compression (DRC), 158–159

E

Eargle, John, 72, 90, 100
Early reflections, 30, 32
Echo, 33, 79, 92
Eighth-sphere radiator, 33
Electrical summation, 25–26
Elementary stream, 153
Emergency service, 153
Emission coders, 149
Encryption, 232–233
Envelopment, of sound system, 28, 37, 50, 109, 187–188
Equal-loudness effect, 159
Equalization, 27, 47, 86, 120–121, 149, 182, 183
Equalizers, 208
  method of setting, 65
Exit sign effect, 47, 116

F

F-weighted, 223
Fantasia, 3, 16, 19, 46
Fantasound, 4
Figure-8. *See* Crossed Figure-8
Film mix, 58
  aux sends, 136
  level adjustment, 162–163
  mixlevel, 160
  PEC/direct switching, 125
Film sound, 8, 10, 42, 46, 56, 65, 164, 182, 197
Five-Point-one channel. *See* 5.1-channel
Fletcher–Munson curves, 55
Floating reference level, 154
Focus control, 114
Folding frequency, 208, 211–212

Frequency masking, 13, 206
Frequency response, 44, 45, 64–67, 81, 89, 117, 120, 132, 182, 206
  measurements, 211
  standardized response, 65–67
Front–back confusion, 180, 189
Fukada array, 72, 94, 135
Full-range monitoring, 25–26, 45

G

Gain staging, 68
Ganged volume control, 125
Gerzon, Michael, 18, 85, 96, 191
Group delay, 130, 204, 205

H

Hamasaki square, 72, 73, 93, 94, 103
Hancock, Herbie, 109, 137–138
HD DVD, 16, 144, 169, 170
Head-related transfer function (HTRF), 7, 41, 89, 111, 180, 182
Headphone listening, 189, 193
Hearing Impaired (HI) channel, 153
Hemispheric radiator, 33
High Fidelity, 26
History, 2–22
Holophone, 72, 94
Horn-loaded high-frequency compression drivers, 33

I

INA, 73, 95
"In-band gain", 60, 67
Indiana Jones and the Last Crusade, 13
In-head localization, 189
Intellectual property protection, 143, 232–233
Interaural level difference (ILD), 7, 179
Interaural time difference (ITD), 7, 179
International Telecommunications Union (ITU), 36, 122, 142
Inter-track synchronization, 127–128

J

Jabba conversation, 56
JBL Professional loudspeaker, 129
Joystick, 111–112
Jurassic Park, 165

K

Knobs, 110–111, 113
Kurtz, Gary, 9, 53, 54

L

Laser Disc, 10, 57, 144, 165
Law of the First Wavefront, 111, 114, 184
  See also Precedence effect
Left and right front speakers, 36
Level calibration, 67
Linear pulse code modulation (LPCM), 123, 134, 144, 152, 215
  2-channel stereo, 168
  5.1-channel, 12, 147
  audio coding, 145–146
  dither, 216–217
  dynamic range, 218–219
  oversampling and noise shaping, 222–223
  performance, 220–222
  word length, 215
  See also Delta-sigma
Lip-sync, 53, 92, 163–164
  phasing, 163
Lissajous display, 128
Listening window, 33
Localization mechanisms, 178–180, 187
  effects on 5.1-channel sound system, 182–183
Lossless coders, 147, 230
Loudness meters, 70
Loudspeaker feeds, time adjustment of, 52–53
Low-bit-rate coder, 12, 16, 44, 164, 171
Low-Frequency Enhancement (LFE), 1, 25, 53–64, 181
  bass management, 56–57
  bottom line, 63–64
  in digital film, 56
  in digital television, 57–58
  film roots, 53–54
  headroom, 54–56
  in home theater, 59
  for music, 59–63
Lt/Rt (left total, right total), 10, 122, 133, 137, 162, 164, 167

M

Martinsound, 68
Masking effects, 148, 196
Massenburg, George, 140
Media specifics, 164–165
Meridian Lossless Packing (MLP), 147, 215, 229
Metadata, 143, 144, 151–152
Mezzanine coder, 124, 149
Microphones
  for 5.1-channel recordings, 93–97
  cardioid, 75
  definitions, 72
  directional, 73
  features, 73–75
  main microphone, 136
  multiple, 76
  noise floor, 74
  placement, 77
  spot microphone, 136
  surround technique, 91–93
  usage, 101
  virtual microphones, 105
  See also Spaced omnis; Coincident technique; Near-coincident technique; Panners
Minimum Audible Angle (MAA), 180, 205
Mixlevel, 159–160
Moiré patterns, 42
Monaural, 8, 181
Monitor loudspeakers, 66, 67
  control room monitor loudspeakers, 33
  direct radiating low-frequency drivers, 33
  directivity index (DI), 33
  horn-loaded high-frequency compression drivers, 33
Monitoring, 23
  frequency response, 64–67
  full-range monitoring, 25–26
  level calibration, 67–70
  loudspeaker feeds, time adjustment of, 52–53
  Low-Frequency Enhancement, 53–64
  monitor loudspeakers, 33–35
  multichannel sound, for room acoustics, 28–33
  spatial balance, 26–28
  standardized setup, 36–39
    compromises, 40–45
    variations, 46–52
Motion Picture Association of America (MPAA), 15
M-S stereo, 82–83, 99
Multichannel audio
  delivery, 144
  for room acoustics, 28–33
  spatial balance, 27–28
  for surround sound, 127
Multichannel microphone technique, 71
  2- and 5-channel, simultaneous recording, 100, 102

Multichannel microphone technique (*contd*)
binaural, 85–86
coincident techniques
crossed Figure-8, 80–82
M-S stereo, 82–83
X-Y stereo, 83
combination of methods, 97–100
dynamic range, 103–105
microphones arrays, for 5.1-channel recordings, 93–97
multichannel perspective, 87–88
near-coincident technique, 83–85
pan pot stereo, 76–78
spaced omnis, 78–80
spot miking, 86–87
standard techniques, 88–91
surround microphone setups, 100, 101
technique, 91–93
upmixing stereo, 103
virtual microphones, 105
Multichannel mixing, 107
double-system audio accompanying video, 123
equalizing, 120–121
equipment and monitor systems, requirement for, 129–131
fitting to digital video recorders, 124
inter-track synchronization, 127–128
mechanics, 110
multichannel monitoring electronics, 125
multichannel outboard gear, 125–127
panners, 110–119
postproduction delivery formats, 134–135
postproduction formats, 132–133
program monitoring, 131–132
reference level, for multichannel program, 123–124
routing multichannel, 121–122
source size, increase in, 119–120
surround mixing experience, 135–138
for DVD music videos, 139–140
track layout, 122–123, 133–134
Multichannel outboard gear, 125–127
decorrelators, 127
monaural, 126
reverberators, 127
spatialized sound, 126
Multichannel panning, 110
2-channel format, 112–113
art of panning, 115–117
direct-sound all round approach, 115
direct/ambient approach, 115
error, 118
HRTFs, 117
in live presentations, 117–118
non-standard panning, 117
panning law, 113–115
time-based panning, 117
Multidirectional radiators, pros and cons, 48–49
Multiple streams, 152–153
dialogue service, 153
Visually Impaired (VI) service, 153
Multiple tracks, 142
"Multi-point mono" approach, 109
Murch, Walter, 114
Music Mostly formats
DTS CD, 227
DVD-A, 227–232
Music Producer's Guild of America (MPGA), 122

N

Near-coincident technique, 80, 83–85
Faulkner stereo, 84
ORTF stereo, 83
sphere microphone, 84–85
Neutral monitoring, 24
Night listening, 159
Noise criterion curves (NC), 28, 29
Noise shaping, 223, 232
Nyquist converters, 212

O

Omnidirectional microphones, 79–80, 92, 97
Omnidirectional radiator, 33
Optimized Cardioid Triangle (OCT), 73
Oversampling, 157, 205, 222–223
Oversampling converters. *See* Delta-sigma

P

Pan pot stereo, 72, 76–78, 88, 89, 135
Panners, 110, 113
DAW software, 112
joystick, 111–112
knobs, 110–111
Passband response, 206
Paul, Les, 4
Perceptual coders, 12, 142, 148, 206
Perceptual hearing mechanisms, 178
Phantom images, 6, 17, 52, 89, 95, 131, 132, 205
in quad, 185–187
stereo, 184–185
Phase cancellation, 25
Phasing, 163
Pink noise, 41, 60, 63, 69
Pinna effects, 179–180
Placement equalization, 45
Polyhymnia array, 73
Postproduction formats, 132–133
delivery, 134–135
dialnorm, 135
Power law, 113
Precedence effect, 52, 87, 111, 113, 130, 187
Pre-masking, 206
Pre-ring effect, 205–206
Pro Logic, 161, 167
Program monitoring, 131–132
phantom image, 132
phase flips, 131
Psychoacoustics
auditory virtual reality, 190
auralization, 190–191
beyond 5.1, 191–193
bass management, 180–182
concert hall acoustics, 188
downmix, 188–189
envelopment, 187–188
Law of the First Wavefront, 184
localization mechanisms, 178–180, 182–183, 187
Low-Frequency Enhancement, 181–182
minimum audible angle, 180
phantom image
in quad, 185–187
stereo, 184–185
spaciousness, 187

Q

Quad, 7, 50, 185–187
Quantization distortion, 145, 216
dither, 216–217
Quantization error. *See* Quantization distortion
Quantizer, 145, 216
Quarter-sphere radiator, 33

R

Radio Shack meter, 60, 70
Reconstruction filter, 203–206, 212

Red Book CD, 147, 232
Reel edits, 164
Re-equalization, 67
Reference level, 142, 160, 224–225
for monitoring, 69
for multichannel program, 123–124
−12 dBFS, 123–124
−18 dBFS, 124
−20 dBFS, 123
Resolution. *See* Word length
Resurrection, 2
Reverberation, 26, 31–32, 89, 92, 93, 103, 109, 120, 127, 136–137, 188
Room acoustics, 28
background noise, 28–29
for multichannel sound, 30
sound isolation, 28
standing wave control, 29–30
Room type
control rooms, 160
Hollywood dubbing stage, 160
Rydstrom, Gary, 195–196

S

Sabine spaces, 31
Sample rate, 149–151, 201–211
aliasing, 209
code dependent distortion, 210
doubling, 209
See also Folding frequency
Sanken CU-41, 75
Sanken WMS-5, 95–96
Satellite broadcast, 171
Satellite–subwoofer system, 57, 59, 180
Saturating functions, 18
Saving Private Ryan, 195
contrast, 197
limitations of surrounds, 198–199
movement of sounds, 197–198
orientation, 196–197
overcoming masking effect, 196
Scheiber, Peter, 8
Schoeps PolarFlex, 75
Sea Biscuit, 18
Sel-sync recording, 4
Shane, 114
Signal-to-noise ratio, 10, 68, 97, 104, 181, 217, 218
"Sin–cos" function, 113
Sine wave, 68, 69, 211, 213, 216
SMART (system-managed audio resource technique), 230
SMPTE (Society of Motion Picture and Television Engineers), 9, 11, 56, 65, 122, 123, 134, 142, 164, 218
reference level, 224
time code, 123, 133
Sony Dynamic Digital Sound (SDDS), 11, 165
Source-playback switching, 125
Source timbre, 182
Sources-all-round approach, 88
Spaced omnis, 72, 75, 78–80, 83
Spaciousness, 26, 80, 85, 99, 109, 187
Spatial balance, 26–28
equalization, 26–27
microphones, 26
Spatial Hearing, 18, 178, 192
Spatial sound, 93, 187
Spatialization technique, 119
Sphere-type stereo microphone, 73
Spot miking, 71, 86–87
Square array, 7, 49–50
Standardized setup
for 5.1-channel sound systems
left and right front speakers, 36
subwoofer, 38
surround loudspeakers, 37
compromises, 40
center loudspeaker, 40–42
left and right speakers, 42–44
subwoofer, 44–45
surround loudspeakers, 44
loudspeaker locations, 38–39
variations
close-field monitoring, 50–52
square array, 49–50
surround arrays, 46–47
surround loudspeaker directivity, 48–49
Star Wars, 8, 9, 10, 16, 53, 54
Static panning, 115
Stereo microphone techniques, 72, 91
Stereo synthesizers, 119–120
Stokowski, Leopold, 3
Subwoofer, 38, 44–45, 58, 60, 63, 181, 192
placement, 30
channel. *See* Low-Frequency Enhancement (LFE)
Summing localization, 184
Super Audio CD, 14, 145, 232, 233
See also DVD-A
Superman, 9
Surround arrays, 4, 9, 35, 46–47
in cinemas, 47
Surround loudspeakers, 6, 28, 35, 37, 40, 44, 46, 50, 92, 189
directivity, 48–49
Surround microphone
for direct sound approach, 93
for direct/ambient approach, 91–93
mixing, 135–138
setups, 100, 101
Sweet spot dependent systems, 189
Symphonie Fantastique, 2

T

Television mix, 58, 118, 125, 150
mix level, 160
Temporal masking, 148
Theile, Günther, 50, 59, 95
This is Cinerama, 5, 6
Three-knob panner, 111–112
Threshold of Pain, 221
Todd, Michael, 6
Top and bottom approach, 42
TPDF dither, 217
Track layout, 122–123, 133–134
Trinnov array, 73, 97
True source timbre, 182

U

Universal Disc Format (UDF), 168
Upmixing stereo, 103

V

VHS, 10
Video-based head tracking, 189
Video games, 171–172
Virtual microphones, 105
Voice-over (VO), 153

W

Waller, Fred, 5
Watermarking, 233
We Were Soldiers, 17
William, Michael, 97

Word length, 149–151, 215–224
  digital audio workstations (DAWs), 150
  See also Linear pulse code modulation (LPCM)
Wrappers, 144

X

X curve, 31, 65, 67
X-Y stereo, 83

Y

Young Sherlock Holmes, 11

Z

Zwicker loudness meter, 156

HAVERING COLLEGE OF F & H E
170533